THE ULTIMATE
ENGAA GUIDE

UniAdmissions

Published by *RAR Medical Services Limited*
www.uniadmissions.co.uk
info@uniadmissions.co.uk
Tel: +44 (0) 208 068 0438

ABOUT THE AUTHORS

Madhivanan graduated with a **1st class degree in Engineering** from the University of Cambridge coming in the top 10% of his year and specialising in bioengineering, instrumentation and control and mechanical engineering. In his final year, he completed a project on the automated diagnosis of otitis media and was awarded the 3rd Year Computer-based Project Prize.

He has won gold medals in the UK Mathematics and Chemistry Olympiads. Madhi has worked with UniAdmissions since 2016 and has successfully tutored many students into Oxbridge. In his spare time, Madhivanan enjoys badminton, chess and travelling.

Peter is one of our Oxbridge admissions tutors at *UniAdmissions*. He is currently in his 4th year of studying Physics at St. Catherine's College, Oxford. He has achieved a distinction in each of his first three years and has been awarded an ATV Scholarship.

Next year he is starting a PhD at Imperial College, focussing on plasma atmospheres around comets. Outside of academia, Peter enjoys playing rugby and escaping into the countryside to climb some rocks.

Rohan is the **Director of Operations** at UniAdmissions and is responsible for its technical and commercial arms. He graduated from Gonville and Caius College, Cambridge and is a fully qualified doctor. Over the last five years, he has tutored hundreds of successful Oxbridge and Medical applicants. He has also authored ten books on admissions tests and interviews.

Rohan has taught physiology to undergraduates and interviewed medical school applicants for Cambridge. He has published research on bone physiology and writes education articles for the Independent and Huffington Post. In his spare time, Rohan enjoys playing the piano and table tennis.

THE ULTIMATE ENGAA GUIDE

MADHIVANAN ELANGO

PETER STEPHENSON

ROHAN AGARWAL

UniAdmissions

ENGAA INTENSIVE COURSE

UNIADMISSIONS

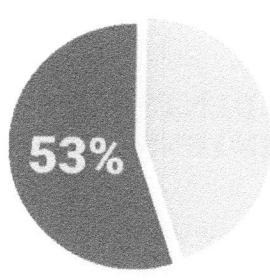

53%

UNIADMISSIONS 3-Year Oxbridge Engineering Programme Success Rate

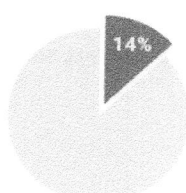

14%

The Average Cambridge Engineering Success Rate

WHY STUDENTS SEE SUCCESS ON OUR INTENSIVE COURSE

1 EXPERT TUTORS.
The course is designed and led by an expert course instructor who has scored in the top 10% of their admissions cycle for this exam. You'll only be taught by the best.

2 GUIDED THROUGH ALL SECTIONS.
You'll be taken through each section of the exam in-depth with a tutor who truly knows the test inside out. They will teach you how to effectively approach each section of the test.

3 LEARN KEY STRATEGIES & TIME-SAVING TIPS.
Throughout the course, you will learn vital strategies to apply when sitting the exam. You will also be taught valuable time-saving tips that help you gain marks most students won't.

300+
Students successfully placed at Oxbridge in the last 3 years

50
Places available on our Oxbridge Engineering Programme

BOOK YOUR **FREE** INTENSIVE COURSE

VISIT: uniadmissions.co.uk/exam-course/

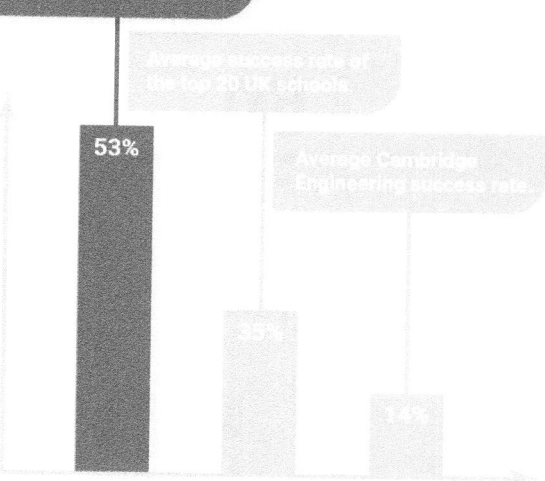

Students enrolled on an Oxbridge Programme.

Average success rate of the top 20 UK schools.

Average Cambridge Engineering success rate

53%

35%

14%

Success Rate

UNIADMISSIONS Oxbridge Engineering Programme Average Success Rate

ENGAA
INTENSIVE COURSE

UNIADMISSIONS

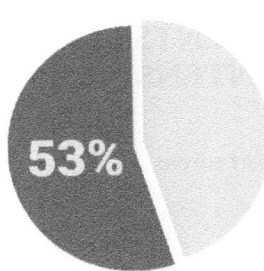

53%

UNIADMISSIONS 3-Year Oxbridge Engineering Programme Success Rate

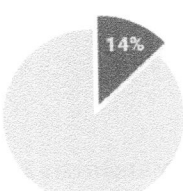

14%

The Average Cambridge Engineering Success Rate

300+

Students successfully placed at Oxbridge in the last 3 years

50

Places available on our Oxbridge Engineering Programme

BOOK YOUR **FREE** INTENSIVE COURSE

VISIT: uniadmissions.co.uk/exam-course/

WHY STUDENTS SEE SUCCESS ON OUR INTENSIVE COURSE

1
EXPERT TUTORS.
The course is designed and led by an expert course instructor who has scored in the top 10% of their admissions cycle for this exam. You'll only be taught by the best.

2
GUIDED THROUGH ALL SECTIONS.
You'll be taken through each section of the exam in-depth with a tutor who truly knows the test inside out. They will teach you how to effectively approach each section of the test.

3
LEARN KEY STRATEGIES & TIME-SAVING TIPS.
Throughout the course, you will learn vital strategies to apply when sitting the exam. You will also be taught valuable time-saving tips that help you gain marks most students won't.

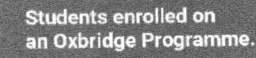

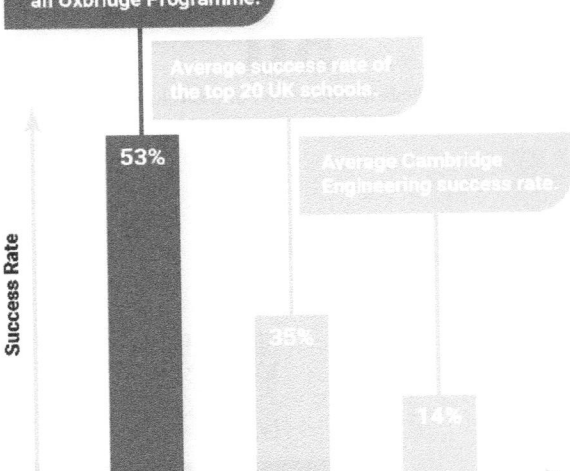

Students enrolled on an Oxbridge Programme.

Average success rate of the top 20 UK schools.

Average Cambridge Engineering success rate.

53%

35%

14%

Success Rate

UNIADMISSIONS Oxbridge Engineering Programme Average Success Rate

THE ULTIMATE ENGAA GUIDE

THE BASICS

What is the ENGAA?

The Engineering Admissions Assessment (ENGAA) is a two-hour written exam taken by prospective Cambridge Engineering applicants.

What does the ENGAA consist of?

Section	Timing	SKILLS TESTED	Questions	Calculator
ONE	60 Minutes	1A: Maths and Physics 1B: Advanced Maths & Advanced Physics	20 MCQs 20 MCQs	Not Allowed
TWO	60 Minutes	Advanced Physics	20 MCQs	Not Allowed

Why is the ENGAA used?

Cambridge Engineering applicants tend to be a bright bunch and usually have excellent grades, with many achieving over 90% in all their A level subjects. This means that competition is fierce – the ENGAA, therefore, is required for universities to differentiate between these top candidates.

When do I sit ENGAA?

The ENGAA normally takes place in the final week of October or the first week of November. As the date varies from year to year, it is best to confirm this on the department's site for prospective students:

https://www.undergraduate.study.cam.ac.uk/courses/engineering#entry-requirements

Can I resit the ENGAA?

No, you can only sit the ENGAA once per admissions cycle.

Where do I sit the ENGAA?

You can usually sit the ENGAA at your school or college (ask your exams officer for more information). Alternatively, if your school is not registered, or you are not attending school, you can sit the ENGAA at an authorised test centre.

Do I have to resit the ENGAA if I reapply?

Yes - you cannot use your score from any previous attempts.

How is the ENGAA scored?

In both sections, each question carries one mark and there is no negative marking.

How is the ENGAA used?

The individual weightings of the ENGAA components vary between Cambridge colleges, so it is important that you email the admissions office, of your chosen college, to understand how your marks will be used. In general, the university will interview a high proportion of realistic applicants, so the ENGAA score is not vital for making the interview shortlist. However, it can play a huge role in the final decision after your interview.

GENERAL ADVICE

Start Early

It is much easier to prepare if you practise in shorter, more frequent sessions. Start your preparation well in advance - ideally by mid-September but at the latest by early October. This will give you plenty of time to complete as many papers as required to help you adequately prepare, and will avoid last-minute panicking and cramming which is a less effective way to learn. In general, an early start will give you the opportunity to identify the complex issues and work at your own pace.

Prioritise

Some questions in the ENGAA can be time-consuming and complex; given the intense time pressure, you need to be aware of your limits. It is essential that you do not get stuck on very difficult questions. If a question looks particularly complicated or involved, mark it for review and move on. You do not want to be caught 5 questions short at the end because you took 3 minutes to solve a multi-step physics question. If a question is taking too long, choose a sensible answer and move on. Remember that each question carries equal weighting and, therefore, you should adjust your timing accordingly. With practice and discipline, you can become increasingly efficient and learn to maximise your performance.

Positive Marking

There are no penalties for incorrect answers in the ENGAA; you will gain one mark for each right answer and miss a mark for each wrong or unanswered question. This provides you with the luxury of resorting to a guess if you run behind on time, or are unable to identify the correct answer. Since each question provides you with 4 to 6 possible answers, you have a 16-25% chance of guessing correctly. Therefore, if you are completely unsure or short on time, then it is best to make an educated guess and move on.

Before simply 'guessing', you should try to eliminate a couple of answers to increase your chances of getting the question correct. For example, if a question has 5 options and you manage to eliminate 2 options, your chances of answering the question correctly increase from 20% to 33%!

Avoid losing easy marks on other questions because of poor exam technique; if you do not manage to finish the exam, take the last 10 seconds to guess the remaining questions to at least give yourself a chance of obtaining more marks.

Practice

This is the best way of familiarising yourself with the style of questions and the timing for this section. Without practise, you are unlikely to be familiar with the style and variety of questions you will encounter in the real examination. Therefore, you want to be comfortable with approaching the style of questions, by adequately practising before you sit the test.

By practising questions, you are less likely to panic during your real examination, which will increase your likelihood of performing well. Initially, work through the questions at your own pace, and spend time carefully reading the questions and looking at any additional data. As you approach the exam date, **make sure you practice the questions under exam conditions.**

Past Papers

The format of the ENGAA has changed for 2019, so past papers do not perfectly reflect the exam you will be sitting this year. Specimen papers are freely available online at **www.uniadmissions.co.uk/ENGAA**. Once you have worked your way through the questions in this book, you are highly advised to attempt them.

Repeat Questions

When checking through answers, pay attention to questions you have answered incorrectly or left blank. If there is a worked answer, follow through the solution carefully, ensuring that you understand the reasoning behind every step, and then repeat the question to check that you can do it independently. If only the answer is given, have another look at the question and attempt to work backwards using the provided answer. This is the best way to learn from your mistakes, and means you are less likely to make similar mistakes when it comes to the test.

The same procedure applies to questions for which you made an educated guess – even if your guess was correct. When working through this book, **make sure you highlight any questions you are unsure of** to indicate you need to spend more time looking over them once marked.

Calculators

You are not permitted to use calculators – thus, it is essential that you have strong numerical skills. For instance, you should be able to rapidly convert between percentages, decimals and fractions. You will seldom get questions that would require a calculator for an exact answer – in this case, you would be expected to arrive at a sensible estimate.

Consider for example:

Estimate 3.962 × 2.322:

3.962 is approximately 4 and 2.323 is approximately 2.33 = 7/3.

Thus, $3.962 \times 2.322 \approx 4 \times \frac{7}{3} = \frac{28}{3} = 9.33$.

Since you will rarely be asked to perform difficult calculations, you can use this as an indicator of whether you are tackling a question correctly. For example, when solving a physics question in section 1, you may end up dividing 8,079 by 357; this should raise alarm bells as calculations in the ENGAA are rarely this difficult.

A word on timing...

"If you had all day to do your ENGAA, you would get 100%. But you don't."

Whilst this is not completely true, it illustrates a very important point. Once you have practiced and know how to approach the style of questions, the clock is your biggest enemy. This seemingly obvious statement has one very important consequence. **The way to improve your ENGAA score is to improve your speed.** There is no magic bullet. But there are a great number of techniques that, with practice, will give you significant time gains, allowing you to answer more questions and score higher marks.

Timing is tight throughout the ENGAA – **mastering timing is the first key to success.** Some candidates choose to work as quickly as possible to allow time at the end to check back, but this is generally not the best method. ENGAA questions can contain a lot of information – each time you start answering a question, it takes time to get familiar with the instructions and information. By splitting the question into two sessions (the first run-through and the return-to-check) you double the amount of time you spend on familiarising yourself with the data, which costs valuable time.

In addition, candidates who do check back may spend 2–3 minutes doing so and yet not make any actual changes. Whilst this can be reassuring, it is a false reassurance as it is unlikely to have a significant effect on your actual score. Therefore, it is usually best to pace yourself very steadily, aiming to spend the same amount of time on each question and finish the final question of the section just as time runs out. This reduces the time spent on re-familiarising with questions and maximises the time spent on the first attempt, gaining more marks.

It is essential that you do not get stuck with the hardest questions – there will undoubtedly be some. In the time spent answering only one of these, you may miss out on answering three easier questions. If a question is taking too long, choose a sensible answer and move on. Never see this as giving up or in any way failing, rather it is the smart way to approach a test with a tight time limit.

With practice and discipline, you can get very good at this and learn to maximise your efficiency. It is not about being a hero and aiming for full marks – this is almost impossible and very much unnecessary (even for Cambridge!). It is about optimising your exam performance and gaining the maximum possible number of marks within the given timeframe.

> *Top tip!* Ensure that you take a watch that can show you the time, in seconds, into the exam. This will allow you have a much more accurate idea of the time you are spending on a question. In general, if you have spent >120 seconds on a section one question, move on regardless of how close you think you are to solving it.

Use the Options:

Some questions may try to overload you with information. When presented with large tables and data, it is essential you look at the answer options so you can focus on the main task. This can allow you to reach the correct answer a lot quicker. Consider the example below:

The table below shows the results of a study investigating antibiotic resistance in staphylococcus populations. A single staphylococcus bacterium is chosen at random from a similar population. Resistance to any one antibiotic is independent of resistance to others.

Antibiotic	Number of Bacteria tested	Number of Resistant Bacteria
Benzyl-penicillin	1011	98
Chloramphenicol	109	1200
Metronidazole	108	256
Erythromycin	105	2

Calculate the probability that the bacterium selected will be resistant to all four drugs.

| 1 in 10^6 | 1 in 10^{20} | 1 in 10^{30} |
| 1 in 10^{12} | 1 in 10^{25} | 1 in 10^{35} |

Studying the options first makes it clear that there is **no need to calculate exact values** – you only need the powers of 10. This makes your life a lot easier. If you had not noticed this, you might have spent well over 90 seconds trying to calculate the exact value when it was not required for the answer.

In other cases, you may be able to use the options to arrive at the solution quicker, by process of elimination. Consider the example below:

A region is defined by the two inequalities: $x - y^2 > 1 \ and \ xy > 1$. Which of the following points is in the defined region?

| (10, 3) | (-10, 3) | (-10, -3) |
| (10, 2) | (-10, 2) | |

Whilst it's possible to solve this question both algebraically or graphically by manipulating the identities, by far **the quickest way is to actually use the options**. Note that options C, D and E violate the second inequality, narrowing down the answers to either A or B. For A: $10 - 3^2 = 1$, which is on the boundary of the defined region and not actually within it. Thus, the answer is B (as $10 - 2^2 = 6$, which satisfies the first inequality).

In general, it pays dividends to inspect the options briefly, to determine whether you can eliminate certain answers. Get into this habit early – it may feel unnatural at first, but it is guaranteed to save you time in the long run.

Keywords

If you are stuck on a question, pay attention to the options that contain key modifiers like "**always**", "**only**", "**all**" as examiners like using them to test if there are any gaps in your knowledge. For example, the statement "arteries carry oxygenated blood" would normally be true; however, "All arteries carry oxygenated blood" would be false because the pulmonary artery carries deoxygenated blood.

SECTION 1: PHYSICS

Section 1 tests your aptitude in Mathematics and Physics. The questions are, approximately, split evenly between GCSE level and A level. You must answer 40 questions in 60 minutes - therefore, this is a very time-pressured section.

The questions can be quite difficult and it is easy to get bogged down. The challenging nature of the questions, coupled with the intense time pressure of having to do one question every 90 seconds, makes this a difficult section.

Gaps in Knowledge

In addition to GCSE level physics and maths, you are also expected to have a firm command of A-level topics. This can be a problem if you have not studied these topics at school yet. A summary of the specification is provided later in the book, but you are highly advised to go through the official ENGAA Specification and ensure that you have covered all examinable topics. An electronic copy of this can be obtained from **www.uniadmissions.co.uk/ENGAA**.

The questions in this book will help highlight any gaps in your knowledge or areas of weakness that you may have. Upon discovering these, make sure you take some time to revise these topics before carrying on – there is little to be gained by attempting questions with huge gaps in your knowledge.

Maths

Mathematical aptitude is extremely important for the ENGAA; many students find that improving their numerical and algebraic skills usually results in notable improvements in their section 1 and 2 scores. Maths pervades the ENGAA – so, if you find yourself consistently running out of time in practice papers, spending a few hours on brushing up your basic maths skills may do wonders for you.

Physics

The syllabus of assumed knowledge, for the physics and advanced physics questions, can be found within the official specification here:

https://www.undergraduate.study.cam.ac.uk/files/publications/engaa_specification_2020.pdf

Multi-Step Questions

Most ENGAA physics questions require two step calculations. Consider the following example:

A metal ball is released from the roof of a 20-metre building. Assuming air resistance is negligible, calculate the velocity at which the ball hits the ground. You are given that $g = 10$ ms^{-2}.

- 5 ms^{-1}
- 10 ms^{-1}
- 15 ms^{-1}
- 20 ms^{-1}
- 25 ms^{-1}

When the ball hits the ground, all of its gravitational potential energy has been converted into kinetic energy.

$$E_p = E_k \;\Rightarrow\; \mathrm{mg\Delta h} = \frac{\mathrm{mv}^2}{2}$$

$$\therefore v = \sqrt{2gh} = \sqrt{2 \times 10 \times 20} = \sqrt{400} = 20\ ms^{-1}$$

In this example, you were required to not only recall two equations, but apply and rearrange them to find a numerical answer – all in under 60 seconds. Note that, if you were comfortable with basic Newtonian mechanics, you could have solved this using a single SUVAT equation:

$$v^2 = u^2 + 2as \;\Rightarrow\; v = \sqrt{2 \times 10 \times 20} = 20\ ms^{-1}$$

SI Units

Remember that, in order to get the correct answer, you must always work in SI units! Do your calculations in terms of metres (not centimetres) and kilograms (not grams), and so on.

Top tip! Knowing SI units is extremely useful because they allow you to **'work out' equations** if you ever forget them. For example, the units for density are kg/m³. Since *kg* is the SI unit for mass, and *m³* is represented by volume, the equation for density must be Density = Mass/Volume.

This can also work the other way; for example, we know that the unit for Pressure is Pascal (Pa). But based on the fact that Pressure = Force/Area, a Pascal must be equivalent to N/m².

Formulas you MUST know:

Equations of Motion (where s = displacement, u = initial velocity, v = final velocity, a = acceleration and t = time):

- $s = ut + \frac{1}{2}at^2$
- $v = u + at$
- $a = \frac{v-u}{t}$
- $v^2 = u^2 + 2as$

Equations Relating to Force:

- Force = Mass x Acceleration (for **constant mass**).
- Force = Rate of change of momentum.
- Pressure = Force / Area.
- Moment of a Force = Force × Distance from pivot.
- Work done = Force x Displacement in direction of force.

Mechanics & Motion:

- Conservation of momentum: $\Delta p = \Delta mv = 0$ (when there is **no net external force**).
- Force = Rate of change of momentum $= \frac{\Delta p}{\Delta t} = \frac{\Delta mv}{\Delta t}$.
- Angular velocity = Rate of change of angular displacement $= \omega = \frac{v}{r} = 2\pi f$ (not explicitly in the ENGAA syllabus but useful for some questions).
- Stress = Force / Area.
- Strain = Extension / Original length.
- Young's modulus = Stress / Strain.

Equations relating to Energy:

- Kinetic Energy $= \frac{1}{2}mv^2$ (for an object with mass **m** and velocity **v**).
- Gravitational Potential Energy $= mgh$ (for an object with mass **m** at height **h** above ground).
- Energy Efficiency = (Useful energy / Total energy) × 100%.

Equations relating to Power:

- Power = Work done / Time.
- Power = Energy transferred / Time.
- Power = Force × Velocity (for **constant force** and velocity **in direction** of the force)

Factor	Text	Symbol
10^{12}	Tera	T
10^9	Giga	G
10^6	Mega	M
10^3	Kilo	k
10^2	Hecto	h
10^{-1}	Deci	d
10^{-2}	Centi	c
10^{-3}	Milli	m
10^{-6}	Micro	μ
10^{-9}	Nano	n
10^{-12}	Pico	p

Magnetic Fields:

- Magnetic force on a **straight, current-carrying wire** in field: $F = BIL$.
- Magnetic flux linkage = Magnetic Flux x Number of coils $= \emptyset N = BAN$.
- Magnitude of induced EMF: $\varepsilon = N\frac{\Delta\emptyset}{\Delta t}$.

Electrical Equations:

- Charge (Q) = Current × Time = It.
- Voltage (V) = Current × Resistance = IR.
- Power = IV = I^{2}R = V^2/R.
- Electromotive Force (EMF) $= \varepsilon = \frac{E}{Q} = I(R + r)$.
- Resistivity $= \rho = \frac{RA}{l}$.
- Resistors in series: $R_T = \sum_{i=1}^{n} R_i$ = **Sum** of all resistances.
- Resistors in parallel: $\frac{1}{R_T} = \sum_{i=1}^{n} \frac{1}{R_i}$ = **Sum** of **reciprocals** of all resistances.

For objects in equilibrium:

- Sum of Clockwise moments = Sum of Anti-clockwise moments.
- Sum of all Resultant Forces = 0.

Waves:

- Speed = Frequency × Wavelength $= c = f\lambda$ (only for **electromagnetic** waves).
- Time Period = 1 / Frequency.
- Snell's law: $n_1 sin\theta_1 = n_2 sin\theta_2$ (where n$_1$, n$_2$ are the refractive indices, θ_1= angle of incidence and θ_2 = angle of refraction).

Radioactivity:

- Decay: $N = N_0 e^{-\lambda t}$ (where N_0 = Initial population size, $\lambda =$ Decay constant).
- Half-life: $T_{1/2} = \frac{ln2}{\lambda}$ (time taken for the population to half).
- Activity: $A = \lambda N$.
- Energy: $E = mc^2$ (where **m** = mass, **c** = speed of light).

Other:

- Weight = Mass × g.
- Density = Mass / Volume.
- Momentum = Mass × Velocity.
- **g = 9.81 ms^{-2}** (unless otherwise stated).

PHYSICS QUESTIONS

Question 1:
Which of the following statements is **FALSE**?

A. Electromagnetic waves can cause substances to heat up.
B. X-rays and gamma rays can knock electrons out of their orbits.
C. Loud sounds can make solid objects vibrate.
D. Wave power can be used to generate electricity.
E. As a wave propagates outwards, the source of the wave loses energy.
F. The amplitude of a wave determines its mass.

Question 2:
A spacecraft is analysing a newly discovered exoplanet. A rock of unknown mass falls, from rest, on the planet from a height of 30 m. Given that $g = 5.4$ ms^{-2} on the planet, calculate the speed of the rock when it hits the ground and the time it took to fall.

	Speed (ms-1)	Time (s)
A	18	3.3
B	18	3.1
C	12	3.3
D	10	3.7
E	9	2.3
F	1	0.3

Question 3:
A canoe floating on the sea rises and falls 7 times in 49 seconds. The waves pass it at a speed of 5 ms^{-1}. How long are the waves?

A. 12 m   C. 25 m   E. 57 m
B. 22 m   D. 35 m   F. 75 m

Question 4:
Miss Orrell lifts her 37.5 kg bike vertically for a distance of 1.3 m in 5 s. The acceleration of free fall is 10 ms^{-2}. What is the average power that she develops?

A. 9.8 W   C. 57.9 W   E. 97.5W
B. 12.9 W   D. 79.5 W   F. 100.0 W

Question 5:

A truck accelerates at 5.6 ms^{-2}, from rest, for 8 seconds. Calculate the final speed and the distance travelled in 8 seconds.

	Final Speed (ms-1)	Distance (m)
A	40.8	119.2
B	40.8	129.6
C	42.8	179.2
D	44.1	139.2
E	44.1	179.7
F	44.2	129.2
G	44.8	179.2
H	44.8	199.7

Question 6:

Which of the following statements is true when a sky diver jumps out of a plane?

A. The sky diver will accelerate until the air resistance is greater than their weight.
B. The sky diver will accelerate until the air resistance is less than their weight.
C. The sky diver will accelerate until the air resistance equals their weight.
D. The sky diver will accelerate until the air resistance equals their weight squared.
E. The sky diver will travel at a constant velocity after leaving the plane.

Question 7:

A 100 g apple falls on Isaac's head from a height of 20 m. Calculate the apple's momentum before the point of impact. Take g = 10 ms^{-2}.

A. 0.2 kgms^{-1}
B. 0.5 kgms^{-1}
C. 1 kgms^{-1}
D. 2 kgms^{-1}
E. 10 kgms^{-1}
F. 20 kgms^{-1}

Question 8:

Which of the following properties do all electromagnetic waves all have in common?

1. They can travel through a vacuum.
2. They can be reflected.
3. They have the same wavelength.
4. They have the same amount of energy.
5. They can be polarised.

A. 1, 2 and 3 only
B. 1, 2, 3 and 4 only
C. 4 and 5 only
D. 3 and 4 only
E. 1, 2 and 5 only
F. 1 and 5 only

Question 9:

A battery with an internal resistance of 0.8 Ω and an EMF of 36 V is used to power a drill with resistance 1 Ω. What is the current in the circuit when the drill is connected to the power supply?

A. 5 A   C. 15 A   E. 25 A

B. 10 A   D. 20 A   F. 30 A

Question 10:

Officer Bailey throws a 20 g dart at a speed of 100 ms^{-1}. It strikes the dartboard and is brought to rest in 10 milliseconds. Calculate the average force exerted on the dart by the dartboard.

A. 0.2 N   C. 20 N   E. 2,000 N

B. 2.0 N   D. 200 N   F. 20,000 N

Question 11:

Professor Huang lifts a 50 kg bag through a distance of 0.7 m in 3 s. What average power does she develop to 3 significant figures? Take $g = 10$ ms^{-2}.

A. 112 W   C. 114 W   E. 116 W

B. 113 W   D. 115 W   F. 117 W

Question 12:

An electric scooter is travelling at a constant speed of 30 ms^{-1}. Its constant speed is maintained by a driving force of 300 N in the direction of motion, which works against a frictional force. Given that the engine runs at 200 V, calculate the current in the scooter, given that the engine is running at 100% efficiency.

A. 4.5 A   C. 450 A   E. 45,000 A

B. 45 A   D. 4,500 A   F. More information needed.

Question 13:

Which of the following statements about the physical definition of work are correct?

1. Work done $= \frac{\text{Force}}{\text{distance}}$.
2. The unit of work is equivalent to kg ms^{-2}.
3. Work is defined as a force multiplied by displacement of the body in the direction of the force.

A. Only 1   D. 1 and 2

B. Only 2   E. 2 and 3

C. Only 3   F. 1 and 3

Question 14:

Which of the following statements about kinetic energy are correct?

1. It is defined as $E_k = \frac{mv^2}{2}$, where m = mass and v = velocity.
2. The unit of kinetic energy is equivalent to Pa × m³.
3. Kinetic energy is equal to the minimum amount of energy needed to decelerate the body in question from its current speed to rest.

A. Only 1
B. Only 2
C. Only 3

D. 1 and 2
E. 2 and 3
F. 1, 2 and 3

Question 15:

In relation to radiation, which of the following statements is **FALSE**?

A. Radiation is the emission of energy in the form of waves or particles.
B. Radiation can be either ionizing or non-ionizing.
C. Gamma radiation has very high energy.
D. Alpha radiation is of lower energy than beta radiation.
E. X-rays are an example of particle radiation.

Question 16:

In relation to the physical definition of half-life, which of the following statements are correct?

1. In radioactive decay, the half-life is independent of atom type and isotope.
2. Half-life is defined as the time required for exactly half of the population to decay.
3. A mass with a constant half-life will exhibit exponential decay.

A. Only 1
B. Only 2
C. Only 3

D. 1 and 2
E. 2 and 3
F. 1 and 3

Question 17:

In relation to nuclear fusion, which of the following statements is **FALSE**?

A. Nuclear fusion is initiated by the absorption of neutrons.
B. Nuclear fusion describes the fusion of hydrogen atoms to form helium atoms.
C. Nuclear fusion releases great amounts of energy.
D. Nuclear fusion requires high activation temperatures.
E. Nuclear fusion is used in stars.

Question 18:

In relation to nuclear fission, which of the following statements is correct?

A. Nuclear fission is the basis of many nuclear weapons.
B. Nuclear fission is triggered by the shooting of neutrons at unstable atoms.
C. Nuclear fission can trigger chain reactions.
D. Nuclear fission commonly results in the emission of ionizing radiation.
E. All of the above.

Question 19:

Two identical resistors (R_a and R_b) are connected in a series circuit. Which of the following statements are true?

1. The current through both resistors is the same.
2. The voltage across both resistors is the same.
3. The voltage across the two resistors is given by Ohm's Law.

A. Only 1
B. Only 2
C. Only 3
D. 1 and 2

E. 2 and 3
F. 1 and 3
G. 1, 2 and 3
H. None of the statements are true.

Question 20:

The Sun is 8 light-minutes away from the Earth. Estimate the circumference of the Earth's orbit around the Sun. Assume that the Earth is in a circular orbit around the Sun. You may take the speed of light to be 3×10^8 ms^{-1}.

A. 10^{24} m
B. 10^{21} m

C. 10^{18} m
D. 10^{15} m

E. 10^{12} m
F. 10^9 m

Question 21:

Which of the following statements are true?

1. Speed is the same as velocity.
2. The standard (SI) unit for speed is ms^{-2}.
3. Velocity = Distance / Time.

A. Only 1
B. Only 2
C. Only 3
D. 1 and 2

E. 2 and 3
F. 1 and 3
G. 1, 2 and 3
H. None of the statements are true.

Question 22:

Which of the following statements best defines Ohm's Law?

A. The current through an insulator between two points is indirectly proportional to the potential difference across the two points.

B. The current through an insulator between two points is directly proportional to the potential difference across the two points.

C. The current through a conductor between two points is inversely proportional to the potential difference across the two points.

D. The current through a conductor between two points is proportional to the square of the potential difference across the two points.

E. The current through a conductor between two points is directly proportional to the potential difference across the two points.

Question 23:

Which of the following statements regarding Newton's Second Law are always correct?

1. For objects at rest, the resultant force acting upon them must be zero.
2. Force = Mass × Acceleration.
3. Force = Rate of change of Momentum.

A. Only 1
B. Only 2
C. Only 3
D. 1 and 2

E. 2 and 3
F. 1 and 3
G. 1, 2 and 3

Question 24:

Which of the following equations concerning electrical circuits are correct?

1. $\text{Charge} = \dfrac{\text{Voltage x time}}{\text{Resistance}}$

2. $\text{Charge} = \dfrac{\text{Power x time}}{\text{Voltage}}$

3. $\text{Charge} = \dfrac{\text{Current x time}}{\text{Resistance}}$

A. Only 1
B. Only 2
C. Only 3
D. 1 and 2

E. 2 and 3
F. 1 and 3
G. 1, 2 and 3
H. None of the statements are true.

Question 25:

An elevator has a mass of 1,600 kg and is carrying passengers that have a combined mass of 200 kg. A constant frictional force of 4,000 N acts upon the elevator. What force must the motor provide for the elevator to move with an upward acceleration of 1 ms⁻²? You may assume that $g = 10$ ms⁻².

A. 1,190 N
B. 11,900 N
C. 18,000 N
D. 22,000 N
E. 23,800 N

Question 26:

A 1,000 kg car accelerates from rest at 5 ms⁻² for 10 seconds. Then, a constant braking force is applied to bring it to rest within 20 seconds. What distance has the car travelled in total?

A. 125 m
B. 250 m
C. 650 m
D. 750 m
E. 1,200 m
F. More information needed.

Question 27:

An electric heater is connected to 120 V mains by a copper wire which has a resistance of 8 ohms. What is the power of the heater?

A. 90 W
B. 180 W
C. 900 W
D. 1800 W
E. 9,000W
F. 18,000 W
G. More information needed.

Question 28:

In a particle accelerator, electron pulses are accelerated through a potential difference of 40 MV and emerge with an energy of 40 MeV (1 MeV = 1.60 × 10⁻¹³ J). Each pulse contains 5,000 electrons. Assuming that the electrons have zero energy prior to being accelerated, what is the power delivered by the electron beam?

A. 1 kW
B. 10 kW
C. 100 kW
D. 1,000 kW
E. 10,000 kW
F. More information needed.

Question 29:

Which of the following statements regarding heat processes is correct?

A. When an object is in equilibrium with its surroundings, there is no energy transferred to or from the object and so its temperature remains constant.
B. When an object is in equilibrium with its surroundings, it radiates and absorbs energy at the same rate and so its temperature remains constant.
C. Radiation is faster than convection but slower than conduction.
D. Radiation is faster than conduction but slower than convection.
E. None of the above.

Question 30:
A 6 kg block is pulled from rest along a horizontal, frictionless surface by a constant, horizontal force of 12 N. Calculate the speed of the block after it has moved 300 cm.

A. $2\sqrt{3}$ ms^{-1}
B. $4\sqrt{3}$ ms^{-1}
C. $4\sqrt{3}$ ms^{-1}
D. 12 ms^{-1}
E. $\sqrt{\frac{3}{2}}$ ms^{-1}

Question 31:
A 100 V heater heats 1.5 litres of pure water from 10°C to 50°C in 50 minutes. Given that 1 kg of pure water requires 4,000 J to raise its temperature by 1°C, calculate the resistance of the heater.

A. 12.5 ohms
B. 25 ohms
C. 125 ohms
D. 250 ohms
E. 500 ohms
F. 850 ohms

Question 32:
Which of the following statements are true?

1. Nuclear fission is the basis of nuclear energy.
2. Following fission, the resulting atoms are a different element to the original type.
3. Nuclear fission often results in the production of free neutrons and photons.

A. Only 1
B. Only 2
C. Only 3
D. 1 and 2
E. 2 and 3
F. 1 and 3
G. 1, 2 and 3
H. None of the statements are true.

Question 33:
Which of the following statements are correct? You may assume that g = 10 ms^{-2}.

- Gravitational potential energy is defined as $\Delta E_p = m \times g \times \Delta h$.
- Gravitational potential energy is a measure of the work done against gravity.
- A reservoir situated 1 km above ground level with 10^6 litres of water has a potential energy of 1 Giga Joule, in reference to ground level.

A. Only 1
B. Only 2
C. Only 3
D. 1 and 2
E. 2 and 3
F. 1 and 3
G. 1, 2 and 3
H. None of the statements are true.

Question 34:
Which of the following statements are correct in relation to Newton's 3rd law?

1. For every action, there is an equal and opposite reaction.
2. According to Newton's 3rd law, there are no isolated forces.
3. Rockets cannot accelerate in deep space because there is no matter to generate an equal and opposite force.

A. Only 1
B. Only 2
C. Only 3
D. 1 and 2
E. 2 and 3
F. 1 and 3

Question 35:

Which of the following statements regarding electrostatic charge are correct?

1. Positively charged objects have gained electrons.
2. The amount of electrical charge flow in a circuit, over a known period of time, can be calculated if the voltage and resistance of the circuit are known.
3. Objects can be charged by friction.

A. Only 1
B. Only 2
C. Only 3
D. 1 and 2

E. 2 and 3
F. 1 and 3
G. 1, 2 and 3

Question 36:

Which of the following statements regarding gravity is true?

A. The gravitational force between two objects is independent of their mass.
B. Each planet in the solar system exerts a gravitational force on the Earth.
C. For satellites in a geostationary orbit, acceleration due to gravity is equal and opposite to the lift from engines.
D. Two objects that are dropped from the Eiffel tower will always land on the ground at the same time if they have the same mass.
E. All of the above.
F. None of the above.

Question 37:

Which of the following best defines an electrical conductor?

A. Conductors are usually made from metals and they conduct electrical charge in multiple directions.
B. Conductors are usually made from non-metals and they conduct electrical charge in multiple directions.
C. Conductors are usually made from metals and they conduct electrical charge in one fixed direction.
D. Conductors are usually made from non-metals and they conduct electrical charge in one fixed direction.
E. Conductors allow the passage of electrical charge with zero resistance because they contain freely mobile charged particles.
F. Conductors allow the passage of electrical charge with maximal resistance because they contain charged particles that are fixed and static.

Question 38:

An 800 kg compact car delivers 20% of its power output to its wheels. If the car has a mileage of 30 miles/gallon and travels at a speed of 60 miles/hour, how much power is delivered to the wheels? 1 gallon of petrol contains 9×10^8 J.

A. 10 kW
B. 20 kW

C. 40 kW
D. 50 kW

E. 100 kW

Question 39:

Which of the following statements regarding beta radiation are true?

1. After a beta particle is emitted, the atomic mass number remains unchanged.
2. Beta radiation can penetrate paper but not aluminium foil.
3. A moving beta particle is deflected in both electric and magnetic fields.

          A. 1 only   C. 1 and 3   E. 2 and 3

          B. 2 only   D. 1 and 2   F. 1, 2 and 3

Question 40:

A car with a weight of 15 kN is travelling at a speed of 15 ms^{-1}. It then crashes into a wall and is brought to rest in 10 milliseconds. Calculate the average braking force exerted on the car by the wall. Take $g = 10$ ms^{-2}.

    A. 1.25×10^4N   C. 1.25×10^6N   E. 2.25×10^5N

    B. 1.25×10^5N   D. 2.25×10^4N   F. 2.25×10^6N

Question 41:

Which of the following statements are correct?

1. Electrical insulators are usually metals, such as copper or aluminium.
2. The flow of charge through electrical insulators is extremely low.
3. Electrical insulators can be charged by rubbing them together.

    A. Only 1   D. 1 and 2   G. 1, 2 and 3

    B. Only 2   E. 2 and 3

    C. Only 3   F. 1 and 3

The following information is needed for Questions 42 and 43:

The graph below represents a car's motion. At t = 0, the car's displacement is zero.

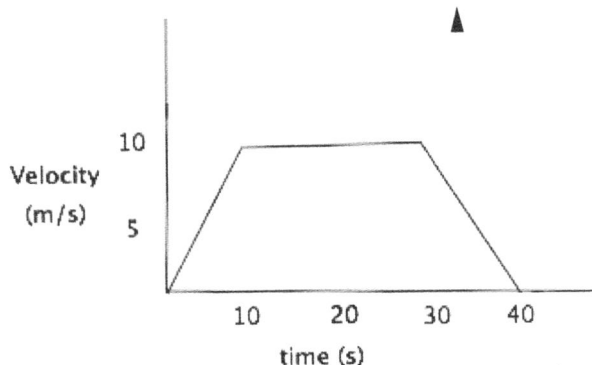

Question 42:

Which of the following statements are **incorrect**?

1. The car is reversing after t = 30 seconds.
2. The car moves with constant acceleration from t = 0 to t = 10 seconds.
3. The car moves with constant speed from t = 10 to t = 30 seconds.

A. Only 1
B. Only 2
C. Only 3
D. 1 and 2

E. 2 and 3
F. 1 and 3
G. 1, 2 and 3

Question 43:

Calculate the total distance travelled by the car during its motion.

A. 200 m
B. 300 m
C. 350 m

D. 400 m
E. 500 m
F. More information needed.

Question 44:

A 1,000 kg rocket is launched and reaches a constant velocity in 30 seconds. Suddenly, a strong gust of wind acts on the rocket for 5 seconds with a force of 10,000 N in its direction of motion. Assuming that the rocket's engines provide a constant force throughout the launch, what is the resulting change in velocity?

A. 0.5 ms^{-1}
B. 5 ms^{-1}
C. 50 ms^{-1}

D. 500 ms^{-1}
E. 5000 ms^{-1}
F. More information needed.

Question 45:

A 0.5 tonne crane lifts a car, with a weight of 0.01 tonnes, by 100 cm vertically in 5,000 milliseconds. Calculate the average power developed by the crane. Take g = 10 ms^{-2}.

A. 0.2 W
B. 2 W
C. 5 W

D. 20 W
E. 50 W
F. More information needed

Question 46:

A 20 V battery is connected to a circuit consisting of a 1 Ω and 2 Ω resistor in parallel. Calculate the total current of the circuit.

A. 6.67 A
B. 8 A

C. 10 A
D. 12 A

E. 20 A
F. 30 A

Question 47:

Which of the following statements is correct?

A. The speed of light changes when it enters water.
B. The speed of light changes when it leaves water.
C. The direction of light changes when it enters water.
D. The direction of light changes when it leaves water.
E. All of the above.
F. None of the above.

Question 48:

In a parallel circuit, a 60 V battery is connected to two branches. Branch A contains 6 identical 5 Ω resistors and branch B contains 2 identical 10 Ω resistors.

Calculate the current in branches A and B.

	I_A (A)	I_B (A)
A	0	6
B	6	0
C	2	3
D	3	2
E	3	3
F	1	5
G	5	1

Question 49:

Calculate the voltage of an electrical circuit that has a power output of 50,000,000,000 nW and a current of 0.000000004 GA.

A. 0.0125 GV
B. 0.0125 MV
C. 0.0125 kV
D. 0.0125 V

E. 0.0125 mV
F. 0.0125 µV
G. 0.0125 nV

Question 50:

Which of the following statements regarding radioactive decay is correct?

A. Radioactive decay of an individual atom is highly predictable.
B. An unstable element will continue to decay until it reaches a stable nuclear configuration.
C. All forms of radioactive decay release gamma rays.
D. All forms of radioactive decay release X-rays.
E. An atom's nuclear charge is unchanged after it undergoes alpha decay.
F. None of the above.

Question 51:

A circuit contains three identical resistors of unknown resistance connected in series with a 15 V battery. The power output of the circuit is 60 W.

Calculate the overall resistance of the circuit when two further identical resistors are added to it in series.

A. 0.125 Ω
B. 1.25 Ω
C.   Ω
D. 6.25 Ω
E. 18.75 Ω
F. More information needed.

Question 52:

A 5,000 kg tractor's engine uses 1 litre of fuel to move 0.1 km. 1 ml of the fuel contains 20 kJ of energy.

Calculate the engine's efficiency. Take $g = 10$ ms^{-2}.

A. 2.5 %
B. 25 %
C. 38 %
D. 50 %
E. 75 %
F. More information needed.

Question 53:

Which of the following statements are correct?

1. Electromagnetic induction occurs when a current-carrying wire moves relative to a magnet.
2. Electromagnetic induction occurs when a magnetic field changes.
3. An electrical current is generated when a coil rotates in a magnetic field.

A. Only 1
B. Only 2
C. Only 3
D. 1 and 2
E. 2 and 3
F. 1 and 3
G. 1, 2 and 3

Question 54:

Which of the following statements are correct regarding parallel circuits?

1. The current flowing through a branch is dependent on the branch's resistance.
2. The total current flowing into the branches is equal to the total current flowing out of the branches.
3. An ammeter will always give the same reading regardless of its location in the circuit.

A. Only 1
B. Only 2
C. Only 3
D. 1 and 2
E. 2 and 3
F. 1 and 3
G. All of the above

Question 55:

Which of the following statements regarding series circuits are true?

1. The overall resistance of a circuit is given by the sum of all resistors in the circuit.
2. Conventional electrical current moves from the positive terminal to the negative terminal.
3. Electrons move from the positive terminal to the negative terminal.

A. Only 1
B. Only 2
C. Only 3

D. 1 and 2
E. 2 and 3
F. 1 and 3

Question 56:

The graphs below show current vs. voltage plots for 4 different electrical components.

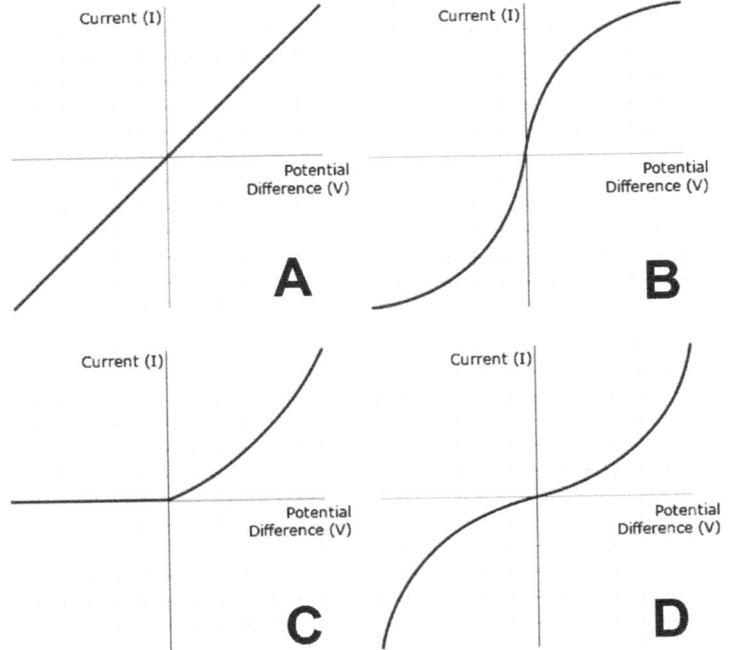

Which of the following graphs represents a resistor at constant temperature, and which a filament lamp?

	Fixed Resistor	Filament Lamp
A	A	B
B	A	C
C	A	D
D	C	A
E	C	C
F	C	D

Question 57:

Which of the following statements are true about vectors?

A. Vectors can be added or subtracted.
B. All vector quantities have a defined magnitude.
C. All vector quantities have a defined direction.
D. Displacement is an example of a vector quantity.
E. All of the above.
F. None of the above.

Question 58:

The acceleration due to gravity on the Earth is six times greater than that on the moon. Dr Tyson records the weight of a rock as 250 N on the moon.

Calculate the rock's density given that it has a volume of 250 cm³. Take g_{Earth} = 10 ms^{-2}.

A. 0.2 kg/cm³   C. 0.6 kg/cm³   E. 0.8 kg/cm³
B. 0.5 kg/cm³   D. 0.7 kg/cm³   F. More information needed.

Question 59:

A radioactive element X_{78}^{225} undergoes alpha decay. What is the atomic mass and atomic number after 5 alpha particles have been released?

	Mass Number	Atomic Number
A	200	56
B	200	58
C	205	64
D	205	68
E	215	58
F	215	73
G	225	78
H	225	83

Question 60:

A 20 A current passes through a circuit with resistance of 10 Ω. The circuit is connected to a transformer that contains a primary coil with 5 turns and a secondary coil with 10 turns. Calculate the potential difference exiting the transformer. You may use the following formula:

$$\frac{\text{Voltage in secondary coil}}{\text{Voltage in primary coil}} = \frac{\text{Turns in secondary coil}}{\text{Turns in primary coil}}$$

A. 100 V   D. 500 V   G. 5,000 V
B. 200 V   E. 2,000 V
C. 400 V   F. 4,000 V

Question 61:

A metal sphere of unknown mass is dropped from an altitude of 1 km and reaches terminal velocity 300 m before it hits the ground. Given that resistive forces do a total of 10 kJ of work for the last 100 m before the ball hits the ground, calculate the mass of the ball. Take $g = 10ms^{-2}$.

A. 1 kg
B. 2 kg
C. 5 kg
D. 10 kg
E. 20 kg
F. More information needed.

Question 62:

Which of the following statements is true about the electromagnetic spectrum?

A. The wavelength of ultraviolet radiation is shorter than that of x-rays.
B. For waves in the electromagnetic spectrum, wavelength is directly proportional to frequency.
C. Most electromagnetic waves can be stopped with a thin layer of aluminium.
D. Waves in the electromagnetic spectrum travel at the speed of sound.
E. Humans are able to visualise the majority of the electromagnetic spectrum.
F. None of the above.

Question 63:

In relation to the Doppler Effect, which of the following statements are true?

1. If an object emitting a wave moves towards the sensor, the wavelength increases and frequency decreases.
2. An object that originally emitted a wave of a wavelength of 20 mm, followed by a second reading delivering a wavelength of 15 mm, is moving towards the sensor.
3. The faster the object is moving away from the sensor, the greater the increase in frequency.

A. Only 1
B. Only 2
C. Only 3
D. 1 and 2
E. 1 and 3
F. 2 and 3
G. 1, 2 and 3
H. None of the above statements are true.

Question 64:

A 5 g bullet travels at 1 km/s and hits a brick wall. It penetrates 50 cm, before being brought to rest, 100 ms after impact. Calculate the average braking force exerted by the wall on the bullet.

A. 50 N
B. 500 N
C. 5,000 N
D. 50,000 N
E. 500,000 N
F. More information needed.

Question 65:

Polonium (Po) is a highly radioactive element that has no known stable isotope. Po^{210} undergoes radioactive decay to Pb^{206} and Y. Calculate the number of protons in 10 moles of Y. [Avogadro's Constant = 6×10^{23}].

A. 0
B. 1.2×10^{24}
C. 1.2×10^{25}
D. 2.4×10^{24}
E. 2.4×10^{25}
F. More information needed

Question 66:

Dr Sale measures a spike of 16,000 Bq from a nuclear rod composed of an unknown material. 300 days later, he visits and can no longer detect a reading higher than 1,000 Bq from the rod, even though the sample has not been disturbed.

What is the longest possible half-life of the nuclear rod?

A. 25 days
B. 50 days
C. 75 days
D. 100 days
E. 150 days
F. More information needed

Question 67:

A radioactive element Y^{200}_{89} undergoes a series of alpha and gamma decays. What are the number of protons and neutrons in the element after the emission of 3 alpha particles and 2 gamma waves?

	Protons	Neutrons
A	79	101
B	83	105
C	83	115
D	89	111
E	89	105
F	93	111
G	93	105
H	109	111

Question 68:

Most symphony orchestras tune to 'standard pitch' (frequency = 440 Hz). When they are tuning, sound directly from the orchestra reaches audience members that are 500 m away in 1.5 seconds.

Estimate the wavelength of 'standard pitch'.

A. 0.05 m
B. 0.5 m
C. 0.75 m
D. 1.5 m
E. 15 m
F. More information needed

Question 69:

A 1 kg cylindrical artillery shell with a radius of 50 mm is fired at a speed of 200 ms⁻¹. It strikes an armour-plated wall and is brought to rest in 500 μs.

Estimate the average pressure exerted on the entire shell, by the wall, at the time of impact.

A. 5×10^6 Pa   C. 5×10^8 Pa   E. 5×10^{10} Pa

B. 5×10^7 Pa   D. 5×10^9 Pa   F. More information needed

Question 70:

A 1,000 W display fountain launches 120 litres of water straight up every minute. Given that the fountain is 10% efficient, calculate the maximum possible height that the stream of water could reach.

Assume that there is negligible air resistance and g = 10 ms⁻².

A. 1 m   C. 10 m   E. 50 m

B. 5 m   D. 20 m   F. More information needed.

Question 71

In relation to transformers, which of the following statements are true?

1. Step-up transformers produce a greater voltage leaving the transformer, compared to the entering voltage.
2. In step-down transformers, the number of turns in the primary coil is smaller than in the secondary coil.
3. For transformers that are 100% efficient: $I_p \times V_p = I_s \times V_s$.

A. Only 1   E. 1 and 3

B. Only 2   F. 2 and 3

C. Only 3   G. 1, 2 and 3

D. 1 and 2   H. None of the above.

Question 72:

The half-life of Carbon-14 is 5,730 years. A bone is found that contains 6.25% of the amount of C^{14} that would be found in a modern one. How old is the bone?

A. 11,460 years   C. 22,920 years   E. 34,380 years

B. 17,190 years   D. 28,650 years   F. 40,110 years

Question 73:

A wave has a velocity of 2,000 mms⁻¹ and a wavelength of 250 cm. What is its frequency in MHz?

A. 8×10^{-3} MHz   C. 8×10^{-5} MHz   E. 8×10^{-7} MHz

B. 8×10^{-4} MHz   D. 8×10^{-6} MHz   F. 8×10^{-8} MHz

Question 74:

A radioactive element has a half-life of 25 days. After 350 days, it has a count rate of 50. What was its original count rate?

A. 102,400

B. 162,240

C. 204,800

D. 409,600

E. 819,200

F. 1,638,400

G. 3,276,800

Question 75:

Which of the following units is **NOT** equivalent to a Volt (V)?

A. $A\Omega$

B. WA^{-1}

C. $Nms^{-1}A^{-1}$

D. NmC

E. JC^{-1}

F. $JA^{-1}s^{-1}$

ADVANCED PHYSICS QUESTIONS

Question 76:
A ball is swung in a vertical circle from a string, of negligible mass, as shown in the diagram.

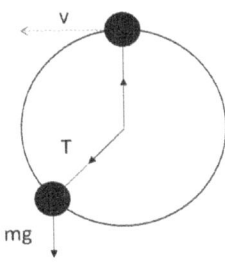

What is the minimum speed at the top of the arc for it to continue in a circular path? You may use the formula $a = \frac{v^2}{r}$, where a is the acceleration of a body undergoing circular motion with velocity v and radius r. The acceleration is directed towards the centre of the circle.

A. 0

B. mgr

C. $2r^2$

D. mg

E. $\sqrt{gr}$

Question 77:
A person pulls on a rope, at 60° to the horizontal, to exert a force on a mass m, as shown. What is the power needed to move the mass up the 30° incline at a constant velocity, v, given a friction force F?

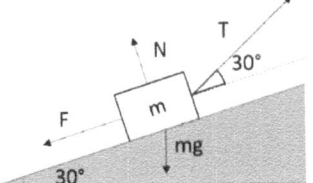

A. $\left(mg + \frac{F}{2}\right)v$

B. $\frac{mg}{\sqrt{2}} - F$

C. $\left(\frac{mg}{2}\right)v$

D. $\sqrt{2}Fv$

E. $\left(\frac{mg}{2} + F\right)v$

Question 78:
What is the maximum speed of a point mass, m, which is suspended from a pendulum of length l, and released from an angle θ?

A. $2gl(1 - \cos(\theta))$

B. $2gl(1 - \sin(\theta))$

C. $\sqrt{2gl(1 - \cos(\theta))}$

D. $\sqrt{2gl(1 - \sin(\theta))}$

E. $\sqrt{2gl(1 - \cos^2(\theta))}$

Question 79:
What would happen to V_{out} if the light intensity upon the circuit below is increased?

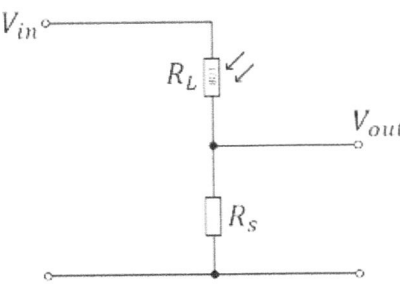

A. Go up

B. Go down

C. Stay the same

D. Decrease to zero

E. Increase infinitely

Question 80:

In the diagram below, the first ball is three times the mass of the other balls. If this is an elastic collision (no loss of kinetic energy), how many of the other balls move and at which velocity after the collision?

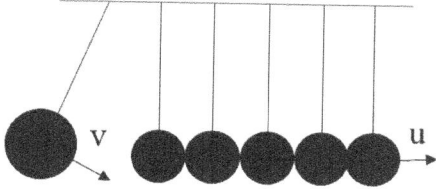

	Number of balls	Velocity
A	1	3v
B	1	v/3
C	3	v/3
D	3	v
E	3	$\sqrt{v}$

Question 81:

A ball is thrown into the air at a velocity 5ms⁻¹ directly upwards and is caught on its descent. Assuming g = 10m/s² and negligible air resistance, what is the displacement of the ball and the distance travelled by the ball after being caught?

A. 0 m, 2.5 m

B. 5 m, 0 m

C. 5 m, 5 m

D. 2.5 m, 0 m

E. 0 m, 2.5 m

F. 0 m, 10 m

Question 82:

The mechanism below is used to weigh two ridged uniform blocks of dimensions 2 metres by 1 metre, and mass 20 kg, on a lever connected to a weighing scale by a cable. What will the reading on the scale be?

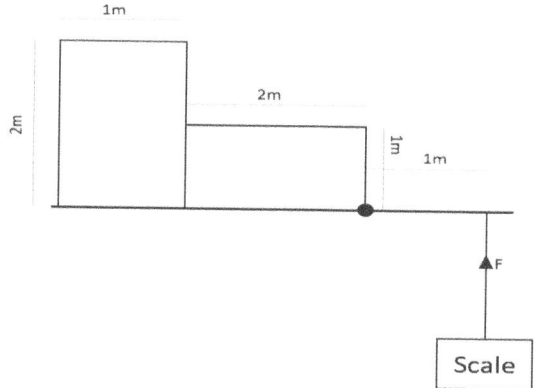

A. 35kg

B. 40kg

C. 70kg

D. 80kg

Question 83:

A ball, of mass m, is dropped from 3 m above the ground and rebounds to a maximum height of 1 m. How much kinetic energy does it have just before hitting the ground and at the top of its bounce, and what is the maximum speed the ball reaches?

	E_k at Bottom	E_k at Top	Max Speed
A	0	30m	$2\sqrt{15}$
B	0	30m	30
C	30m	0	$2\sqrt{15}$
D	30m	0	60
E	60m	0	60

Question 84:

Which of the following beams could not be in equlibirum, regardless of the magnitudes of the forces (assuming they are not zero)? The arrows in the diagrams below represent forces.

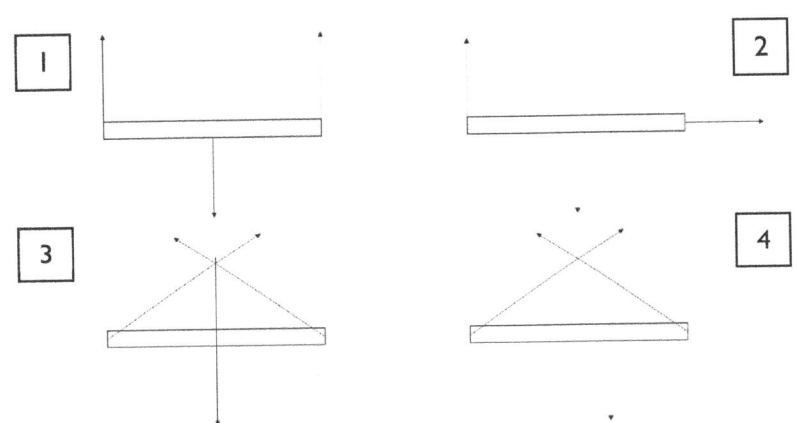

A. 1 and 2   C. 2 and 4   E. Only 4

B. 1 and 3   D. Only 1   F. 3 and 4

Question 85:

What is the stopping distance of a car, moving at a velocity v, if its braking force is half its weight?

A. v^2 C. 2mv   E. $\sqrt{mg}$

B. $\dfrac{v^2}{g}$ D. $\dfrac{v^2}{2}$

Question 86:

The amplitude of a wave is damped from an initial amplitude of 200 to 25 over 12 seconds. How many seconds did it take to reach half its original amplitude? Assume that the wave undergoes exponential decay.

A. 1
B. 2
C. 3
D. 4
E. 6

Question 87:

A block is sliding at a speed v with an acceleration a along a rough ground. Which of the following expressions represents the power dissipated, assuming the friction coeffiecient is μ and its mass is m?

A. amg
B. μmgv
C. μamg
D. μmv

Question 88:

Study the diagram provided. The two waves can represent:

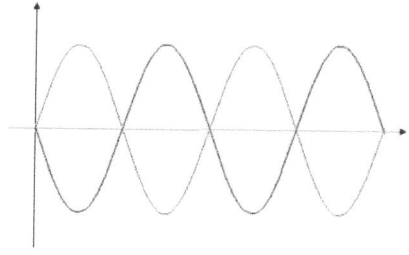

A. A standing wave with both ends fixed.
B. The 4th harmonic.
C. Destructive interference.
D. A reflection from a plane surface.
E. All of the above.

Question 89:

Radioactive element $_b^a$X undergoes beta decay, in which a neutron is changed into a proton, and the product of this decay emits an alpha particle to become $_d^c$Y. What are the atomic number and atomic mass?

	c	d
A	a - 4	b + 1
B	a - 3	b - 2
C	a - 4	b - 1
D	a - 5	b
E	a - 1	b - 4

Question 90:

Consider a spherical shell of radius r and thickness t, where r >> t (r is significantly greater than t). If the inside is pressurised to p, above atmospheric pressure, what is the stress in the walls of the sphere?

A. pr/2t
B. 2rtπ
C. 4πr³/3
D. pr/t

Question 91:

For the arrangement of springs below, what is the spring constant of the whole system, assuming each spring has spring constant k?

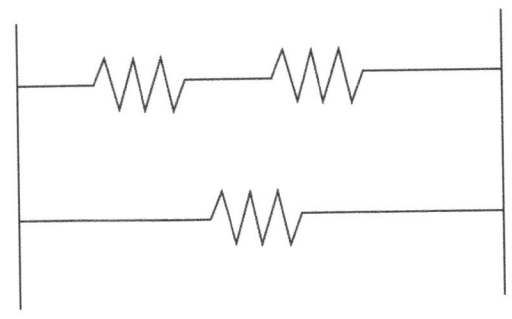

A. $k/2$ B. k C. $3k/2$ D. $2k^2$ E. $2k/3$

Question 92:

Consider the arrangement below showing a car dragging a trailer using a connection of stiffness 100000 N/m. At an instant they are accelerating at 2 m/s². Assuming the car weighs 20 Mg and the trailer weighs 10 Mg, what is the energy stored in the spring?

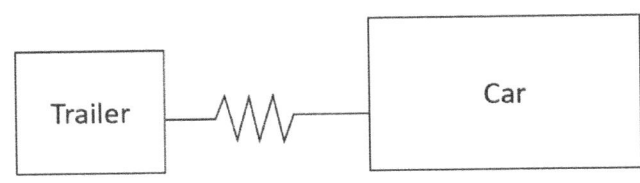

A. 2000J   B. 2J   C. 2000kJ   D. 8000J

Question 93:

A blue LED has a power output of 10 W, and is connected to a source with an EMF of 5V. Estimate the number of electrons passing through the LED in 10 seconds and the energy provided by the battery to each individual electron.

	Number of electrons	Energy of an electron (J)
A	1.2×10^{19}	2×10^{-19}
B	2×10^{19}	8×10^{-19}
C	1.25×10^{20}	8×10^{-19}
D	1.5×10^{20}	12×10^{-19}
E	3×10^{25}	4×10^{-19}

Question 94:

A flower pot hangs on the end of a rod protruding at right angles from a wall, held up by string attached two thirds of the way along from the wall. What must the tension in the string be if the rod is weightless and the system is at equilibrium?

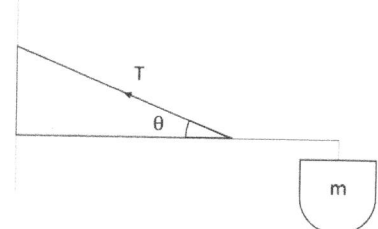

A. $mg \sin \theta$

B. $\frac{3mg}{2\sin \theta}$

C. $\frac{3mg}{2\cos \theta}$

D. $\frac{2mg}{3\sin \theta}$

E. $\frac{2mg}{3\cos \theta}$

Question 95:

Consider two signals: a 5V signal at 30kHz and a 10V signal at 50kHz. What is the time period of the combined signal?

A. 6.67 µs

B. 150 µs

C. 6.67 ms

D. 80 ms

E. 80 µs

Question 96:

For the graph of a material below estimate E, the Young's Modulus, and estimate the strain energy at point x, the elastic limit.

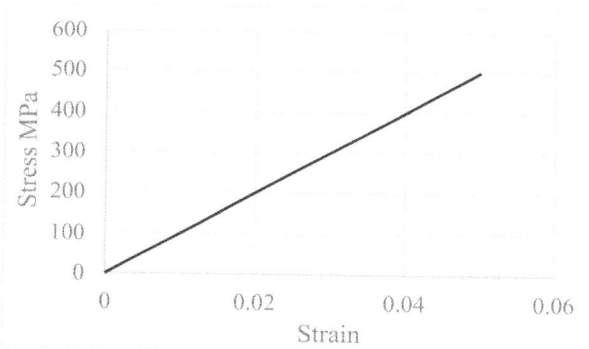

A. 210 GPa and 10000 kJ.

B. 10 GPa and 12.5 MJ.

C. 210 GPa and 6.125 MJ.

D. 10 GPa and 10000 kJ.

Question 97:

A pool cue gives a ball of mass 100g an impulse of 0.27 Ns. What is the velocity of the ball after the impact?

A. 27 m/s

B. 0.027 ms

C. 0.27 m/s

D. 2.7 m/s

E. 3.6 m/s

Question 98:

A ball is thrown at velocity v. Assuming negligible air resistance, and that the ball forms a projectile path, what is the optimal angle to throw the ball to get the furthest distance?

A. 10

B. 30

C. 45

D. 60

E. 80

Question 99:

An electric car travels at constant speed v, with a motor of efficiency 90% and consumes electrical power P. What is the work done by air resistance in a distance of 1 km?

A. 900v

B. 900P/v

C. 0.9P/v

D. 900v/P

E. 90P/v

Question 100:

A block of mass m is on a smooth wedge. The wedge accelerates at a in the horizontal direction. What acceleration a is required from the wedge so that the block stays still? The wedge makes an angle θ to the horizontal.

A. mg

B. $g \tan \theta$

C. $g \cos \theta$

D. $mg \sin \theta$

E. m^2g

SECTION 1: MATHS

The syllabus for the maths and advanced maths section of the ENGAA can be found within the official specification. A link is provided here:
https://www.undergraduate.study.cam.ac.uk/files/publications/engaa_specification2019.pdf.

Core Formulas:

2D Shapes			3D Shapes	
Area			Surface Area	Volume
Circle	πr^2	Cuboid	Σ of 6 faces	Length × Width × Height
Parallelogram	Base × Vertical height	Cylinder	$2\pi r^2 + 2\pi rl$	$\pi r^2 \times h$
Trapezium	0.5 × Vertical height × (a + b)	Cone	$\pi r^2 + \pi rl$	$\pi r^2 \times (h/3)$
Triangle	0.5 × Base × Height	Sphere	$4\pi r^2$	$(4/3)\pi r^3$

Even good students who are studying maths at A2 can struggle with certain ENGAA maths topics because they're usually glossed over at school. These include:

Quadratic Formula

The solutions for a quadratic equation in the form $ax^2 + bx + c = 0$ are given by: $x = \frac{-b \pm \sqrt{b^2 - 4ac}}{2a}$.

Remember that you can also use the discriminant to quickly determine whether a quadratic equation has any solutions:

$b^2 - 4ac < 0 \Rightarrow$ No solutions

$b^2 - 4ac = 0 \Rightarrow$ 1 repeated solution

$b^2 - 4ac > 0 \Rightarrow$ 2 distinct solutions

Completing the Square

If a quadratic equation cannot be factorised easily, and is in the format $ax^2 + bx + c = 0$, then you can rearrange it into the form $a\left(x + \frac{b}{2a}\right)^2 \left[c - \frac{b^2}{4a}\right] = 0$

This looks more complicated than it is – remember that in the ENGAA, you're extremely unlikely to get quadratic equations where $a > 1$ or an equation which doesn't have any easy factors. This gives you an easier equation: $\left(x + \frac{b}{2}\right)^2 + \left[c - \frac{b^2}{4}\right] = 0$ and is best understood with an example.

Consider: $x^2 + 6x + 10 = 0$.

This equation cannot be factorised easily but note that: $x^2 + 6x - 10 = (x + 3)^2 - 19 = 0$.

Therefore, $x = -3 \pm \sqrt{19}$. Completing the square is an important skill – make sure you're comfortable with it.

Difference between 2 Squares

If you are asked to simplify expressions and find that there are no common factors, but the expression does involve square numbers, you might be able to factorise it by using the 'difference between two squares' trick.

For example, $x^2 - 25$ can also be expressed as$(x + 5)(x - 5)$.

MATHS QUESTIONS

Question 101:

Robert has a box of building blocks. The box contains 8 yellow blocks and 12 red blocks. He picks three blocks from the box and stacks them up high. Calculate the probability that he stacks two red building blocks and one yellow building block, in **any** order.

A. $\frac{8}{20}$

B. $\frac{44}{95}$

C. $\frac{11}{18}$

D. $\frac{8}{19}$

E. $\frac{12}{20}$

F. $\frac{35}{60}$

Question 102:

Solve $\frac{3x+5}{5} + \frac{2x-2}{3} = 18$.

A. 12.11

B. 13.49

C. 13.95

D. 14.2

E. 19

F. 265

Question 103:

Solve $3x^2 + 11x - 20 = 0$.

A. 0.75 and $-\frac{4}{3}$

B. -0.75 and $\frac{4}{3}$

C. -5 and $\frac{4}{3}$

D. 5 and $\frac{4}{3}$

E. 12 only

F. -12 only

Question 104:

Express $\frac{5}{x+2} + \frac{3}{x-4}$ as a single fraction.

A. $\frac{15x-120}{(x+2)(x-4)}$

B. $\frac{8x-26}{(x+2)(x-4)}$

C. $\frac{8x-14}{(x+2)(x-4)}$

D. $\frac{15}{8x}$

E. 24

F. $\frac{8x-14}{x^2-8}$

Question 105:

The value of p is directly proportional to the cube root of q. When p = 12, q = 27. Find the value of q when p = 24.

A. 32

B. 64

C. 124

D. 128

E. 216

F. 1728

Question 106:

Write 72^2 as a product of its prime factors.

A. $2^6 \times 3^4$

B. $2^6 \times 3^5$

C. $2^4 \times 3^4$

D. 2×3^3

E. $2^6 \times 3$

F. $2^3 \times 3^2$

Question 107:

Calculate: $\dfrac{2.302 \times 10^5 + 2.302 \times 10^2}{1.151 \times 10^{10}}$.

A. 0.0000202

B. 0.00020002

C. 0.00002002

D. 0.00000002

E. 0.000002002

F. 0.000002002

Question 108:

Given that $y^2 + ay + b = (y + 2)^2 - 5$, find the values of **a** and **b**.

	a	b
A	-1	4
B	1	9
C	-1	-9
D	-9	1
E	4	-1
F	4	1

Question 109:

Express $\dfrac{4}{5} + \dfrac{m-2n}{m+4n}$ as a single fraction in its simplest form.

A. $\dfrac{6m+6n}{5(m+4n)}$

B. $\dfrac{9m+26n}{5(m+4n)}$

C. $\dfrac{20m+6n}{5(m+4n)}$

D. $\dfrac{3m+9n}{5(m+4n)}$

E. $\dfrac{3(3m+2n)}{5(m+4n)}$

F. $\dfrac{6m+6n}{3(m+4n)}$

Question 110:

A is inversely proportional to the square root of B. When A = 4, B = 25.

Calculate the value of A when B = 16.

A. 0.8

B. 4

C. 5

D. 6

E. 10

F. 20

Question 111:

S, T, U and V are points on the circumference of a circle, as shown in the diagram, and O is the centre of the circle.

Given that angle SVU = 89°, calculate the size of the smaller angle SOU.

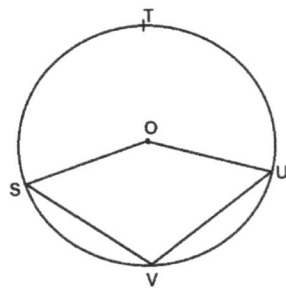

A. 89°

B. 91°

C. 102°

D. 178°

E. 182°

F. 212°

Question 112:

Open cylinder A has a surface area of 8π cm² and a volume of 2π cm³. Open cylinder B is an enlargement of A and has a surface area of 32π cm². Calculate the volume of cylinder B.

A. 2π cm³

B. 8π cm³

C. 10π cm³

D. 14π cm³

E. 16π cm³

F. 32π cm³

Question 113:

Express $\frac{8}{x(3-x)} - \frac{6}{x}$ in its simplest form.

A. $\frac{3x-10}{x(3-x)}$

B. $\frac{3x+10}{x(3-x)}$

C. $\frac{6x-10}{x(3-2x)}$

D. $\frac{6x-10}{x(3+2x)}$

E. $\frac{6x-10}{x(3-x)}$

F. $\frac{6x+10}{x(3-x)}$

Question 114:

A bag contains 10 balls, where 9 are white and 1 is black. What is the probability that the black ball is drawn in the tenth and final draw if the drawn balls are not replaced?

A. 0

B. $\frac{1}{10}$

C. $\frac{1}{100}$

D. $\frac{1}{10^{10}}$

E. $\frac{1}{362,880}$

Question 115:

Gambit has an ordinary deck of 52 cards. What is the probability of Gambit drawing 2 Kings (without replacement)?

A. 0

B. $\frac{1}{169}$

C. $\frac{1}{221}$

D. $\frac{4}{663}$

E. None of the above.

Question 116:

There are two identical unfair dice, where the probability that the dice gets a 6 is twice as high as the probability of any other outcome, which are all equally likely. What is the probability that rolling both dice will give a total will be 12?

A. 0

B. $\frac{4}{49}$

C. $\frac{1}{9}$

D. $\frac{2}{7}$

E. None of the above.

Question 117:

A roulette wheel consists of 36 numbered spots and 1 zero spot (i.e. 37 spots in total).

What is the probability that the ball will stop in a spot either divisible by 3 or 2?

A. 0

B. $\frac{25}{37}$

C. $\frac{25}{36}$

D. $\frac{18}{37}$

E. $\frac{24}{37}$

Question 118:

An unbiased coin is flipped 4 times. What is the probability of obtaining 2 heads and 2 tails?

A. $\frac{1}{16}$

C. $\frac{3}{8}$

E. None of the above.

B. $\frac{3}{16}$

D. $\frac{9}{16}$

Question 119:

Shivun rolls two fair dice. What is the probability that he obtains a total of 5, 6 or 7?

A. $\frac{9}{36}$

C. $\frac{1}{6}$

E. None of the above.

B. $\frac{7}{12}$

D. $\frac{5}{12}$

Question 120:

Dr Savary has a bag that contains x red balls, y blue balls and z green balls. He pulls out a ball, replaces it, and then pulls out another. What is the probability that he picks one red ball and one green ball?

A. $\frac{2(x+y)}{x+y+z}$

C. $\frac{2xz}{(x+y+z)^2}$

E. $\frac{4xz}{(x+y+z)^4}$

B. $\frac{xz}{(x+y+z)^2}$

D. $\frac{(x+z)}{(x+y+z)^2}$

F. More information needed.

Question 121:

Mr Kilbane has a bag that contains x red balls, y blue balls and z green balls. He pulls out a ball, does **NOT** replace it, and then pulls out another. What is the probability that he picks one red ball and one blue ball?

A. $\frac{2xy}{(x+y+z)^2}$

C. $\frac{2xy}{(x+y+z)^2}$

E. $\frac{4xy}{(x+y+z-1)^2}$

B. $\frac{2xy}{(x+y+z)(x+y+z-1)}$

D. $\frac{xy}{(x+y+z)(x+y+z-1)}$

F. More information needed.

Question 122:

There are two tennis players. The first player wins the point with probability p, and the second player wins the point with probability 1- p. The first player to score four points wins the game, unless the score is 4 - 3. At this point, the first player to get two points ahead wins.

What is the probability that the first player wins in exactly 5 rounds?

A. $4p^4(1\text{-}p)$

C. $4p(1\text{-}p)$

E. $4p^5(1\text{-}p)$

B. $p^4(1\text{-}p)$

D. $4p(1\text{-}p)^4$

F. More information needed.

Question 123:

Solve the equation $\frac{4x + 7}{2} + 9x + 10 = 7$.

A. $\frac{22}{13}$

C. $\frac{10}{13}$

E. $\frac{13}{22}$

B. $-\frac{22}{13}$

D. $-\frac{10}{13}$

F. $-\frac{13}{22}$

Question 124:

The volume of a sphere is $V = \frac{4}{3}\pi r^3$, and the surface area of a sphere is $S = 4\pi r^2$. Express S in terms of V.

A. $S = (4\pi)^{2/3}(3V)^{2/3}$ C. $S = (4\pi)^{1/3}(9V)^{2/3}$ E. $S = (16\pi)^{1/3}(9V)^{2/}$

B. $S = (8\pi)^{1/3}(3V)^{2/3}$ D. $S = (4\pi)^{1/3}(3V)^{2/3}$

Question 125:

Express the volume of a cube, V, in terms of its surface area, S.

A. $V = (S/6)^{3/2}$ C. $V = (6/S)^{3/2}$ E. $V = (S/36)^{1/2}$

B. $V = S^{3/2}$ D. $V = (S/6)^{1/2}$ F. $V = (S/36)^{3/2}$

Question 126:

Solve the equations $4x + 3y = 7$ and $2x + 8y = 12$.

A. $(x,y) = \left(\frac{17}{13}, \frac{10}{13}\right)$ D. $(x,y) = (2,1)$ G. No solutions possible.

B. $(x,y) = \left(\frac{10}{13}, \frac{17}{13}\right)$ E. $(x,y) = (6,3)$

C. $(x,y) = (1,2)$ F. $(x,y) = (3,6)$

Question 127:

Rearrange $\frac{(7x+10)}{(9x+5)} = 3y^2 + 2$, to make x the subject.

A. $\dfrac{15\,y^2}{7 - 9(3y^2+2)}$ C. $-\dfrac{15\,y^2}{7 - 9(3y^2+2)}$ E. $-\dfrac{5\,y^2}{7 + 9(3y^2+2)}$

B. $\dfrac{15\,y^2}{7 + 9(3y^2+2)}$ D. $-\dfrac{15\,y^2}{7 + 9(3y^2+2)}$ F. $\dfrac{5\,y^2}{7 + 9(3y^2+2)}$

Question 128:

Simplify $3x\left(\dfrac{3x^7}{x^{\frac{1}{3}}}\right)^3$.

A. $9x^{20}$ C. $87x^{20}$ E. $27x^{21}$

B. $27x^{20}$ D. $9x^{21}$ F. $81x^{21}$

Question 129:

Simplify $2x[(2x)^7]^{\frac{1}{14}}$.

A. $2x\sqrt{2\,x^4}$ C. $2\sqrt{2\,x^4}$ E. $8x^3$

B. $2x\sqrt{2x^3}$ D. $2\sqrt{2x^3}$ F. $8x$

Question 130:
What is the circumference of a circle with an area of 10π?

A. $2\pi\sqrt{10}$ C. 10π E. $\sqrt{10}$

B. $\pi\sqrt{10}$ D. 20π F. More information needed.

Question 131:
Evaluate the value of $(3.4).5$, given that $a\,.\,b = (ab) + (a + b)$.

A. 19   C. 100   E. 132

B. 54   D. 119

Question 132:
Calculate $(2.3).2$, given that $a.\,b = \dfrac{a^b}{a}$.

A. $\dfrac{16}{3}$ C. 2   E. 8

B. 1   D. 4

Question 133:
Solve $x^2 + 3x - 5 = 0$.

A. $x = -\dfrac{3}{2} \pm \dfrac{\sqrt{11}}{2}$ C. $x = -\dfrac{3}{2} \pm \dfrac{\sqrt{11}}{4}$ E. $x = \dfrac{3}{2} \pm \dfrac{\sqrt{29}}{2}$

B. $x = \dfrac{3}{2} \pm \dfrac{\sqrt{11}}{2}$ D. $x = \dfrac{3}{2} \pm \dfrac{\sqrt{11}}{4}$ F. $x = -\dfrac{3}{2} \pm \dfrac{\sqrt{29}}{2}$

Question 134:
How many times do the curves $y = x^3$ and $y = x^2 + 4x + 14$ intersect?

A. 0   C. 2   E. 4

B. 1   D. 3

Question 135:
Which of the following graphs **do not** intersect?

1. $y = x$
2. $y = x^2$
3. $y = 1 - x^2$
4. $y = 2$

A. 1 and 2   C. 3 and 4   E. 1 and 4

B. 2 and 3   D. 1 and 3   F. 2 and 4

Question 136:
Calculate the product of 897,653 and 0.009764.

A. 87646.8   C. 876.468   E. 8.76468
B. 8764.68   D. 87.6468   F. 0.876468

Question 137:
Solve for x: $\frac{7x+3}{10} + \frac{3x+1}{7} = 14$.

A. $\frac{929}{51}$ C. $\frac{949}{79}$
B. $\frac{949}{47}$ D. $\frac{980}{79}$

Question 138:
What is the area of an equilateral triangle with side length x.

A. $\frac{x^2\sqrt{3}}{4}$ C. $\frac{x^2}{2}$ E. x^2
B. $\frac{x\sqrt{3}}{4}$ D. $\frac{x}{2}$ F. x

Question 139:
Simplify $3 - \frac{7x(25x^2 - 1)}{49x^2(5x+1)}$.

A. $3 - \frac{5x-1}{7x}$ C. $3 + \frac{5x-1}{7x}$ E. $3 - \frac{5x^2}{49}$
B. $3 - \frac{5x+1}{7x}$ D. $3 + \frac{5x+1}{7x}$ F. $3 + \frac{5x^2}{49}$

Question 140:
Solve the equation $x^2 - 10x - 100 = 0$.

A. $-5 \pm 5\sqrt{5}$ C. $5 \pm 5\sqrt{5}$ E. $5 \pm 5\sqrt{125}$
B. $-5 \pm \sqrt{5}$ D. $5 \pm \sqrt{5}$ F. $-5 \pm \sqrt{125}$

Question 141:
Rearrange $x^2 - 4x + 7 = y^3 + 2$ to make x the subject.

A. $x = 2 \pm \sqrt{y^3 + 1}$ C. $x = -2 \pm \sqrt{y^3 - 1}$
B. $x = 2 \pm \sqrt{y^3 - 1}$ D. $x = -2 \pm \sqrt{y^3 + 1}$

Question 142:
Rearrange $3x + 2 = \sqrt{7x^2 + 2x + y}$ to make y the subject.

A. $y = 4x^2 + 8x + 2$ C. $y = 2x^2 + 10x + 2$ E. $y = x^2 + 10x + 2$
B. $y = 4x^2 + 8x + 4$ D. $y = 2x^2 + 10x + 4$ F. $y = x^2 + 10x + 4$

Question 143:

Rearrange $y^4 - 4y^3 + 6y^2 - 4y + 2 = x^5 + 7$ to make y the subject.

A. $y = 1 + (x^5 + 7)^{1/4}$

B. $y = -1 + (x^5 + 7)^{1/4}$

C. $y = 1 + (x^5 + 6)^{1/4}$

D. $y = -1 + (x^5 + 6)^{1/4}$

Question 144:

The aspect ratio of my television screen is 4:3 and the diagonal is 50 inches. What is the area of my television screen?

A. 1,200 inches²

B. 1,000 inches²

C. 120 inches²

D. 100 inches²

E. More information needed.

Question 145:

Rearrange the equation $\sqrt{1 + 3x^{-2}} = y^5 + 1$ to make x the subject.

A. $x = \frac{(y^{10} + 2y^5)}{3}$

B. $x = \frac{3}{(y^{10} + 2y^5)}$

C. $x = \sqrt{\frac{3}{y^{10} + 2y^5}}$

D. $x = \sqrt{\frac{y^{10} + 2y^5}{3}}$

E. $x = \sqrt{\frac{y^{10} + 2y^5 + 2}{3}}$

F. $x = \sqrt{\frac{3}{y^{10} + 2y^5 + 2}}$

Question 146:

Solve $3x - 5y = 10$ and $2x + 2y = 13$.

A. $(x, y) = \left(\frac{19}{16}, \frac{85}{16}\right)$

B. $(x, y) = \left(\frac{85}{16}, -\frac{19}{16}\right)$

C. $(x, y) = \left(\frac{85}{16}, \frac{19}{16}\right)$

D. $(x, y) = \left(-\frac{85}{16}, -\frac{19}{16}\right)$

E. No solutions possible.

Question 147:

The two inequalities $x + y \le 3$ and $x^3 - y^2 < 3$ define a region on a plane. Which of the following points is inside the region?

A. (2, 1)

B. (2.5, 1)

C. (1, 2)

D. (3, 5)

E. (1, 2.5)

F. None of the above.

Question 148:

How many points of intersection do $y = x + 4$ and $y = 4x^2 + 5x + 5$ have?

A. 0

B. 1

C. 2

D. 3

E. 4

Question 149:

How many points of intersection do $y = x^3$ and $y = x$ have?

A. 0

B. 1

C. 2

D. 3

E. 4

Question 150:

A cube has unit length sides. What is the length of a line joining a vertex to the midpoint of the opposite side?

A. $\sqrt{2}$

B. $\sqrt{\dfrac{3}{2}}$

C. $\sqrt{3}$

D. $\sqrt{5}$

E. $\dfrac{\sqrt{5}}{2}$

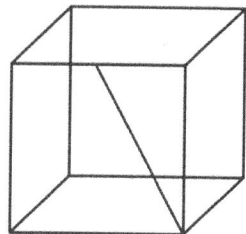

Question 151:

Solve for x, y, and z in the following system of equations.

1. $x + y - z = -1$
2. $2x - 2y + 3z = 8$
3. $2x - y + 2z = 9$

	x	y	z
A	2	-15	-14
B	15	2	14
C	14	15	-2
D	-2	15	14
E	2	-15	14
F	No solutions possible		

Question 152:

Fully factorise the following expression: $3a^3 - 30a^2 + 75a$.

A. $3a(a - 3)^3$

B. $a(3a - 5)^2$

C. $3a(a^2 - 10a + 25)$

D. $3a(a - 5)^2$

E. $3a(a + 5)^2$

Question 153:

Solve for x and y in the following simultaneous equations.

$4x + 3y = 48$

$3x + 2y = 34$

	x	y
A	8	6
B	6	8
C	3	4
D	4	3
E	30	12
F	12	30
G	No solutions possible	

Question 154:

Evaluate: $\dfrac{-\left(5^2 - 4 \times 7\right)^2}{-6^2 + 2 \times 7}$.

A. $-\dfrac{3}{50}$

B. $\dfrac{11}{22}$

C. $-\dfrac{3}{22}$

D. $\dfrac{9}{50}$

E. $\dfrac{9}{22}$

F. 0

Question 155:

All license plates are 6 characters long. The first 3 characters consist of letters and the next 3 characters of numbers. How many unique license plates are possible?

A. 676,000

B. 6,760,000

C. 67,600,000

D. 1,757,600

E. 17,576,000

F. 175,760,000

Question 156:

How many solutions are there for: $2(2(x^2 - 3x)) = -9$.

A. 0

B. 1

C. 2

D. 3

E. Infinite solutions.

Question 157:

Evaluate: $\left(x^{\frac{1}{2}} y^{-3}\right)^{\frac{1}{2}}$.

A. $\dfrac{x^{\frac{1}{2}}}{y}$

B. $\dfrac{x}{y^{\frac{3}{2}}}$

C. $\dfrac{x^{\frac{1}{4}}}{y^{\frac{3}{2}}}$

D. $\dfrac{y^{\frac{1}{4}}}{x^{\frac{3}{2}}}$

Question 158:

Bryan earned a total of £ 1,240 last week from renting out three flats. From this, he had to pay 10% of the rent from the 1-bedroom flat for repairs, 20% of the rent from the 2-bedroom flat for repairs, and 30% from the 3-bedroom flat for repairs. The 3-bedroom flat costs twice as much as the 1-bedroom flat. Given that the total repair bill was £ 276 calculate the rent for each apartment.

	1 Bedroom	2 Bedrooms	3 Bedrooms
A	280	400	560
B	140	200	280
C	420	600	840
D	250	300	500
E	500	600	1,000

Question 159:

Evaluate: $5\left[5(6^2 - 5 \times 3) + 400^{\frac{1}{2}}\right]^{1/3} + 7.$

A. 0

B. 25

C. 32

D. 49

E. 56

F. 200

Question 160:

What is the area of a regular hexagon with side length 1?

A. $3\sqrt{3}$

B. $\frac{3\sqrt{3}}{2}$

C. $\sqrt{3}$

D. $\frac{\sqrt{3}}{2}$

E. 6

F. More information needed

Question 161:

Dexter moves into a new rectangular room that is 19 metres longer than it is wide, and its total area is 780 square metres. What are the room's dimensions?

A. Width = 20 m; Length = -39 m

B. Width = 20 m; Length = 39 m

C. Width = 39 m; Length = 20 m

D. Width = -39 m; Length = 20 m

E. Width = -20 m; Length = 39 m

Question 162:

Tom uses 34 meters of fencing to enclose his rectangular lot. He measured the diagonals to 13 metres long. What is the length and width of the lot?

A. 3 m by 4 m

B. 5 m by 12 m

C. 6 m by 12 m

D. 8 m by 15 m

E. 9 m by 15 m

F. 10 m by 10 m

Question 163:

Solve $\frac{3x-5}{2} + \frac{x+5}{4} = x + 1$.

A. 1
B. 1.5

C. 3
D. 3.5

E. 4.5
F. None of the above

Question 164:

Calculate: $\frac{5.226 \times 10^6 + 5.226 \times 10^5}{1.742 \times 10^{10}}$.

A. 0.033
B. 0.0033

C. 0.00033
D. 0.00003

E. 0.0000033

Question 165:

Calculate the area of the triangle shown to the right:

A. $3 + \sqrt{2}$
B. $\frac{2 + 2\sqrt{2}}{2}$

C. $2 + 5\sqrt{2}$
D. $3 - \sqrt{2}$

E. 3
F. 6

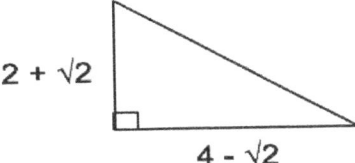

$2 + \sqrt{2}$

$4 - \sqrt{2}$

Question 166:

Rearrange $\sqrt{\frac{4}{x}} + 9 = y - 2$ to make x the subject.

A. $x = \frac{11}{(y-2)^2}$
B. $x = \frac{9}{(y-2)^2}$

C. $x = \frac{4}{(y+1)(y-5)}$
D. $x = \frac{4}{(y-1)(y+5)}$

E. $x = \frac{4}{(y+1)(y+5)}$
F. $x = \frac{4}{(y-1)(y-5)}$

Question 167:

When 5 is subtracted from 5x the result is half the sum of 2 and 6x. What is the value of x?

A. 0
B. 1

C. 2
D. 3

E. 4
F. 6

Question 168:

Estimate $\frac{54.98 + 2.25^2}{\sqrt{905}}$.

A. 0
B. 1

C. 2
D. 3

E. 4
F. 5

Question 169:

At a Pizza Parlour, you can order single, double or triple cheese in the crust. You also have the option to include ham, olives, pepperoni, bell pepper, meat balls, tomato slices, and pineapples. How many different types of pizza are available at the Pizza Parlour?

A. 10

B. 96

C. 192

D. 384

E. 768

F. None of the above.

Question 170:

Solve the simultaneous equations $x^2 + y^2 = 1$ and $x + y = \sqrt{2}$, for x, y > 0.

A. $(x, y) = \left(\frac{\sqrt{2}}{2}, \frac{\sqrt{2}}{2}\right)$

B. $(x, y) = \left(\frac{1}{2}, \frac{\sqrt{3}}{2}\right)$

C. $(x, y) = (\sqrt{2} - 1, 1)$

D. $(x, y) = (\sqrt{2}, \frac{1}{2})$

Question 171:

Which of the following statements is **FALSE**?

A. Congruent objects always have the same dimensions and shape.

B. Congruent objects can be mirror images of each other.

C. Congruent objects do not always have the same angles.

D. Congruent objects can be rotations of each other.

E. Two triangles are congruent if they have two sides and one angle of the same magnitude.

Question 172:

Solve the inequality: $x^2 \geq 6 - x$.

A. $x \leq -3$ and $x \leq 2$

B. $x \leq -3$ and $x \geq 2$

C. $x \geq -3$ and $x \leq 2$

D. $x \geq -3$ and $x \geq 2$

E. $x \geq 2$ only

F. $x \geq -3$ only

Question 173:

The hypotenuse of an isosceles, right-angled triangle is x cm. The other two sides are equal in length. What is the area of the triangle in terms of x?

A. $\frac{\sqrt{x}}{2}$

B. $\frac{x^2}{4}$

C. $\frac{x}{4}$

D. $\frac{3x^2}{4}$

E. $\frac{x^2}{10}$

Question 174:

Mr Heard derives a formula: $Q = \frac{(X+Y)^2 A}{3B}$. He doubles the values of X and Y, halves the value of A and triples the value of B. What happens to value of Q?

A. Decreases by $\frac{1}{3}$ C. Decreases by $\frac{2}{3}$ E. Increases by $\frac{4}{3}$

B. Increases by $\frac{1}{3}$ D. Increases by $\frac{2}{3}$ F. Decreases by $\frac{4}{3}$

Question 175:

Consider the graphs $y = x^2 - 2x + 3$, and $y = x^2 - 6x - 10$. Which of the following is true?

A. Both equations intersect the x-axis.

B. Neither equation intersects the x-axis.

C. The first equation does not intersect the x-axis; the second equation intersects the x-axis.

D. The first equation intersects the x-axis; the second equation does not intersect the x-axis.

ADVANCED MATHS QUESTIONS

Question 176:

The vertex of an equilateral triangle is covered by a circle whose radius is half the height of the triangle. What percentage of the triangle is covered by the circle?

A. 12%
B. 16%
C. 23%
D. 33%
E. 41%
F. 50%

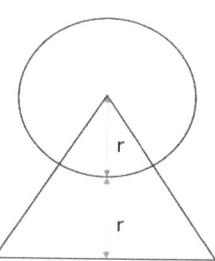

Question 177:

Three identical circles, which each have a radius r, fit into a quadrilateral as shown. Determine the height of the quadrilateral.

A. $2\sqrt{3}r$
B. $(2 + \sqrt{3})r$
C. $(4 - \sqrt{3})r$
D. $3r$
E. $4r$
F. More information needed.

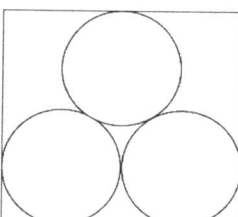

Question 178:

Two pyramids have equal volume and height, one with a square base of side length **a** and one with a hexagonal base of side length **b**. What is the ratio of the side length of the bases?

A. $\sqrt{\dfrac{3\sqrt{3}}{2}}$

B. $\sqrt{\dfrac{2\sqrt{3}}{3}}$

C. $\sqrt{\dfrac{3}{2}}$

D. $\dfrac{2\sqrt{3}}{3}$

E. $\dfrac{3\sqrt{3}}{2}$

Question 179:

One 9 cm cube is cut into 3 cm cubes. The total surface area increases by a factor of:

A. $\dfrac{1}{3}$

B. $\sqrt{3}$

C. 3

D. 9

E. 27

Question 180:
A cone has height twice its base width (four times the circle radius). What is the cone angle (half the angle at the vertex)?

A. $30°$

B. $\sin^{-1}\left(\frac{r}{2}\right)$

C. $\tan^{-1}\left(\frac{1}{4}\right)$

D. $\cos^{-1}(\sqrt{17})$

Question 181:
A hemispherical speedometer has a maximum speed of 200 mph. What is the angle travelled by the needle at a speed of 70 mph?

A. $28°$

B. $49°$

C. $63°$

D. $88°$

E. $92°$

Question 182:
Two rhombuses, A and B, are similar. The area of A is 10 times that of B. What is the ratio of the smallest angles over the ratio of the shortest sides?

A. 0

B. $\frac{1}{10}$

C. $\frac{1}{\sqrt{10}}$

D. $\sqrt{10}$

E. ∞

Question 183:
If $f^{-1}(-x) = \ln(2x^2)$, what is $f(x)$?

A. $\sqrt{\frac{e^y}{2}}$

B. $\sqrt{\frac{e^{-y}}{2}}$

C. $\frac{e^y}{2}$

D. $\frac{-e^y}{2}$

E. $\sqrt{\frac{e^y}{2}}$

Question 184:
Which of the following has the largest value for $0 < x < 1$?

A. $\log_8(x)$

B. $\log_{10}(x)$

C. e^x

D. x^2

E. $\sin(x)$

Question 185:
The variable x is proportional to y cubed; y is proportional to the square root of z.

If z doubles, x changes by a factor of:

A. $\sqrt{2}$

B. 2

C. $2\sqrt{2}$

D. $\sqrt[3]{4}$

E. 4

Question 186:

The area between two concentric circles (shaded) is three times that of the inner circle.

What is the size of the gap?

A. r

B. $\sqrt{2}r$

C. $\sqrt{3}r$

D. $2r$

E. $3r$

F. $4r$

Question 187:

Solve $-x^2 \leq 3x - 4$.

A. $x \geq \frac{4}{3}$

B. $1 \leq x \leq 4$

C. $x \leq 2$

D. $x \geq 1$ or $x \geq -4$

E. $-1 \leq x \leq \frac{3}{4}$

Question 188:

The volume of a sphere is equal to its projected area. What is its radius?

A. $\frac{1}{2}$

B. $\frac{2}{3}$

C. $\frac{3}{4}$

D. $\frac{4}{3}$

E. $\frac{3}{2}$

Question 189:

What is the range where $x^2 < \frac{1}{x}$?

A. $x < 0$

B. $0 < x < 1$

C. $x > 0$

D. $x \geq 1$

E. None.

Question 190:

Simplify and solve the following expression:

$(e - a) (e + b) (e - c) (e + d)...(e - z).$

A. 0

B. e^{26}

C. e^{26} (a-b+c-d...+z)

D. e^{26} (a+b-c+d...-z)

E. e^{26} (abcd...z)

F. None of the above.

Question 191:

Find the value of k such that the vectors $\mathbf{a} = -\mathbf{i} + 6\mathbf{j}$ and $\mathbf{b} = 2\mathbf{i} + k\mathbf{j}$ are perpendicular.

A. -2

B. $-\frac{1}{3}$

C. $\frac{1}{3}$

D. 2

Question 192:

What is the perpendicular distance between point **p** with position vector $4\mathbf{i} + 5\mathbf{j}$ and the line L given by vector equation $\mathbf{r} = -3\mathbf{i} + \mathbf{j} + \lambda(\mathbf{i} + 2\mathbf{j})$.

A. $2\sqrt{7}$ C. $2\sqrt{5}$

B. $5\sqrt{2}$ D. $7\sqrt{2}$

Question 193:

Find k such that point $\begin{pmatrix} 2 \\ k \\ -7 \end{pmatrix}$ lies within the plane $\mathbf{r} = \begin{pmatrix} 2 \\ 3 \\ -1 \end{pmatrix} + \lambda\begin{pmatrix} 4 \\ 1 \\ 0 \end{pmatrix} + \mu\begin{pmatrix} 2 \\ 1 \\ 3 \end{pmatrix}$.

A. -2 C. 0 E. 2

B. -1 D. 1

Question 194:

What is the largest solution to $\sin\left(\frac{\pi}{2} - 2\theta\right) = 0.5$ for $\frac{\pi}{2} \leq x \leq 2\pi$?

A. $\frac{5\pi}{3}$ C. $\frac{5\pi}{6}$ E. $\frac{11\pi}{6}$

B. $\frac{4\pi}{3}$ D. $\frac{7\pi}{6}$

Question 195:

$\cos^4(x) - \sin^4(x) \equiv$

A. $\cos(2x)$ C. $\sin(2x)$ E. $\tan(x)$

B. $2\cos(x)$ D. $\sin(x)\cos(x)$

Question 196:

How many real roots does $y = 2x^5 - 3x^4 + x^3 - 4x^2 - 6x + 4$ have?

A. 1 C. 3 E. 5

B. 2 D. 4

Question 197:

What is the sum of 8 terms, $\sum_1^8 u_n$, of an arithmetic progression with $u_1 = 2$ and $d = 3$?

A. 15 C. 100 E. 282

B. 82 D. 184

Question 198:

What is the coefficient of the x^2 term in the binomial expansion of $(2 - x)^5$?

A. -80 C. 40 E. 80

B. -48 D. 48

Question 199:

Given you have already thrown a 6, what is the probability of throwing three consecutive 6s using a fair die?

A. $\frac{1}{216}$

B. $\frac{1}{36}$

C. $\frac{1}{6}$

D. $\frac{1}{2}$

E. 1

Question 200:

Three people, A, B and C, play darts. The probability that they hit a bull's eye are respectively $\frac{1}{5}, \frac{1}{4}, \frac{1}{3}$. What is the probability that at least two shots hit the bullseye?

A. $\frac{1}{60}$

B. $\frac{1}{30}$

C. $\frac{1}{12}$

D. $\frac{1}{6}$

E. $\frac{3}{20}$

Question 201:

If probability of having blonde hair is 1 in 4, the probability of having brown eyes is 1 in 2, and the probability of having both is 1 in 8, what is the probability of having neither blonde hair nor brown eyes?

A. $\frac{1}{2}$

B. $\frac{3}{4}$

C. $\frac{3}{8}$

D. $\frac{5}{8}$

E. $\frac{7}{8}$

Question 202:

Differentiate and simplify the following expression: $y = x(x+3)^4$.

A. $(x+3)^3$

B. $(x+3)^4$

C. $x(x+3)^3$

D. $(5x+3)(x+3)^3$

E. $5x^3(x+3)$

Question 203:

Evaluate $\int_1^2 \frac{2}{x^2} dx$.

A. -1

B. $\frac{1}{3}$

C. 1

D. $\frac{21}{4}$

E. 2

Question 204:

Express $\frac{5i}{1+2i}$ in the form $a + bi$.

A. $1 + 2i$

B. $4i$

C. $1 - 2i$

D. $2 + i$

E. $5 - i$

Question 205:

Simplify $7\log_a(2) - 3\log_a(12) + 5\log_a(3)$.

A. $\log_{2a}(18)$

B. $\log_a(18)$

C. $\log_a(7)$

D. $9\log_a(17)$

E. $-\log_a(7)$

Question 206:

What is the equation of the asymptote of the function $y = \frac{2x^2 - x + 3}{x^2 + x - 2}$?

A. $x = 0$

B. $x = 2$

C. $y = 0.5$

D. $y = 0$

E. $y = 2$

Question 207:

Find the intersection(s) of the functions $y = e^x - 3$ and $y = 1 - 3e^{-x}$.

A. 0 and $\ln(3)$

B. 1

C. $\ln(4)$ and 1

D. $\ln(3)$

Question 208:

Find the radius of the circle $x^2 + y^2 - 6x + 8y - 12 = 0$.

A. 3

B. $\sqrt{13}$

C. 5

D. $\sqrt{37}$

E. 12

Question 209:

Which value of a minimises the magnitude of $\int_0^a 2\sin(-x)\,dx$?

A. 0.5π

B. π

C. 2π

D. 3π

E. 5π

Question 210:

When $\frac{2x+3}{(x-2)(x-3)^2}$ is expressed as a sum of partial fractions, what is the numerator in the $\frac{A}{(x-2)}$ term?

A. -7

B. -1

C. 3

D. 6

E. 7

SECTION 2

Section 2 of the ENGAA consists of 20 questions. You have 60 minutes to answer all the questions. Section 2 requires the same core knowledge as section 1.

Watch the Clock

The key to maximising your score is to have a strategy of how best to use your time. Each question is split into four parts, but not all parts are worth the same number of marks. Firstly, you should be aware of how long you should ideally be spending on each question on average. With 32 marks available, this equates to around one mark per minute, but some will take longer than others. An easier goal is to think that four questions, of 6-8 marks each, in 40 minutes means you should spend under 10 minutes per question, to (ideally) leave time for reading and checking.

Start with the 'Low Hanging Fruit'

If you spot a question you know immediately how to tackle, that is where you should start. This will quickly add to your marks, provide extra time, and also build your confidence before tackling the more challenging or unfamiliar questions. As the grouped questions follow on from each other, that might mean answering the beginning of all four questions or one question entirely, whichever you are most comfortable with. However, be warned that the more you jump around between different types of question, the more likely you are to make mistakes or waste time reacquainting yourself with the problem at hand. Once you have gone through the paper once and answered the 'obvious' questions, you can then reassess how long you have left and how many questions remain to best allocate your time before moving on.

Practice

The best way to prepare is to practice not only long answer questions of this level, but also different multiple-choice papers, to familiarise yourself with eliminating options and obtaining an answer as fast as possible. The questions in this book, although written in the style prior to the 2018 revision of the ENGAA specification, are a good starting point as the content of the questions and their difficulty is unchanged. As the exam is relatively new, there are not many past papers to use, however there are several exam boards as well as competitions like the Canadian Maths Competition that are multiple choice, which you can use for practice.

Thrive on Adversity

The ENGAA is specifically designed to be challenging and to take you out of your comfort zone. This is done in order to separate different tiers of students depending on their academic ability. The reason for this type of exam is that Cambridge attracts excellent students that will almost invariably score well in exams. If, for this reason, during your preparation, you come across questions that you find very difficult, use this as motivation to try and further your knowledge beyond the simple school syllabus. This is where the option for specialisation ties in.

Practice Calculus

Practicing calculus might seem unnecessary, but confidence in the core calculus techniques will allow you to answer questions significantly faster and with improved accuracy. As the math part is core knowledge for all section 2 questions, you can be certain that to some degree or other you will be required to apply your calculus knowledge.

Think in Applied Formulas

Attempting to answer Section 2 Physics questions without first learning all of the standard formulae is like trying to run before you can walk – ensure you're completely confident with all the core formulas before starting the practice questions.

Variety

Physics is a very varied subject and the questions you may be asked in an exam will reflect this variety. Many students have a preferred area within physics, such as electronics, astrophysics, or mechanics. However, it is important to remember not to neglect any subject area in its entirety. It is entirely possible that of the questions in the ENGAA, several of them could be outside of your comfort zone – leaving you in a difficult position.

Graphs

Graph-sketching is usually a tricky area for many students. When tackling a graph-sketching problem, there are many approaches; however, it is useful to start with the basic features of the plot:

- What is the value of y when x is zero?
- What is the value of x when y is zero?
- Are there any special values of x and y?
- If there is a fraction involved, at what values of x is the numerator or denominator equal to zero?

If you asked to draw a function that is the sum, product or division of two functions, start by drawing out these sub-functions. Which function is the dominant function when x > 0 and x < 0?

Answering these basic questions will tell you where the asymptotes and intercepts are, which will help with drawing the function.

Remember the Basics

A surprisingly large number of students do not know what the properties of basic shapes are. For example, the area of a circle is πr^2, the surface area of a sphere is $4\pi r^2$ and the volume of a sphere is $\frac{4}{3}\pi r^3$. To 'go up' a dimension (i.e. to go from an area to a volume) you need to integrate and to 'go down' a dimension you need to differentiate. Learning this will make it easy to remember formulas for the areas and volumes of basic shapes – the constant at the front, however, will need to be memorised.

It is also important to remember important formulas, even though they may be included in formula booklets. The reason for this is that although you may have access to formula booklets during exams, this will not be the case in interviews which will follow the ENGAA. In addition, flicking through formula booklets takes up time during an exam and can be avoided if you are able to memorise important formulas.

QUESTIONS

Question 1.1

A golfer swings a club so that the head completes a (virtually) complete circle in T = 0.1 s. The length of the club is R = 1 m.

What angle from the horizontal should the club hit the ball at to maximise the distance travelled?

A. 0°

B. 30°

C. 45°

D. 90

E. Any angle

Question 1.2

What is the (approximate) velocity of the club when it strikes the ball?

A. m/s

B. 3 m/

C. 10 m/s

D. 60 m/

E. 100 m/s

Question 1.3

Assuming that the ground is flat, what is the total time the golf ball is in the air?

A. $3\sqrt{2}$ s

B. $5\sqrt{2}$

C. $6\sqrt{2}$ s

D. 12

E. 15 s

Question 1.4

What therefore is the maximum horizontal distance the ball travels?

A. 180 m

B. $180\sqrt{2}$

C. 360 m

D. $360\sqrt{2}$

E. 720 m

Question 2.1

What is the mass of the moon in terms of the radius of the earth, r, if the density of the Moon is 75% of the density of earth, ρ, and that the radius of the Moon is 4 times smaller than the radius of Earth.

A. $\frac{1}{64}\rho\pi r^3$

B. $\frac{9}{16}\rho\pi r^3$

C. $\frac{3}{64}\rho\pi r^2$

D. $\frac{3}{64}\rho\pi r^3$

E. $\frac{1}{16}\rho\pi r^3$

Question 2.2

Given that gravitational force is determined by $F = G\frac{Mm}{R^2}$, which of the following is an expression for the acceleration due to gravity on earth, where G is the gravitational constant?

A. $G\rho\pi r^2$

B. $\frac{3}{4}G\rho\pi r$

C. $\frac{4}{3}G\rho\pi r^2$

D. $\frac{1}{16}G\rho\pi r$

E. $\frac{4}{3}G\rho\pi r$

Question 2.3
Estimate the gravitational acceleration at the surface of the Moon relative to g on earth.

A. $8g$
B. $\frac{g}{8}$
C. $\frac{g}{16}$
D. $\frac{3g}{16}$
E. $\frac{7g}{16}$

Question 2.4
How would the orbital speed of a satellite, of equal mass and equal orbital radius, change between orbiting earth and the moon?

A. Decrease by a factor of $\sqrt{3}/4$.
B. Decrease by a factor of $\frac{7}{16}$.
C. It will be remain unchanged.
D. Increase by a factor of $\frac{1}{16}$.
E. Increase by a factor of 8.

Question 3.1
Identical resistors are connected using wire of negligible resistance to a 1.4 V power supply. What would the resistance be of two resistors in parallel?

A. $2R$
B. $\frac{R}{2}$
C. $\frac{3R}{2}$
D. $\frac{2R}{3}$
E. R

Question 3.2
What would the total resistance in this circuit be, considering all the resistors to be identical?

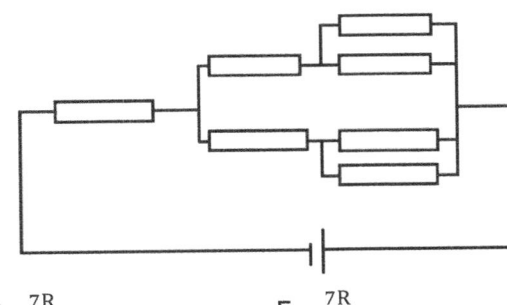

A. $7R$
B. $3R$
C. $\frac{7R}{3}$
D. $\frac{3R}{4}$
E. $\frac{7R}{4}$

Question 3.3
What would R be to produce a total current of 2 A?

A. Ω
B. $0.4\ \Omega$
C. Ω
D. $0.5\ \Omega$
E. Ω

Question 3.4
What would be the power dissipated in the circuit?

A. 0.6 W
B. 1.2 W
C. 2.3 W
D. 2.8 W
E. W

Question 4.1: The position of a particle the moves along the x axis is given by the equation $x = 10 + 1.5t^3$. What expression would represent the velocity of the particle?

A. $v = 1 + 1.5t^2$ C. $v = 10 + 4.5t^2$ E. $v = 3.0t^3$

B. $v = 4.5t^2$ D. $v = 45 + 4.5t^3$

Question 4.2: At which time, t, the acceleration of the particle is equal to zero?

A. $t = 20$ s   C. $t = 4.5$ s   E. $t = 1.5$ s

B. $t = 10$ s   D. $t = 0$

Question 4.3: What is the average velocity of the particle for the time interval t = 2 to 10 seconds?

A. $v = 150$ m/s   C. $v = 190$ m/s   E. $v = 3.0t^3$

B. $v = 400$ m/s   D. $v = 250$ m/s

Question 4.4: If the equation represents the motion of the ball, of mass m, that is thrown against a wall. What will be the energy transferred to the wall in an elastic collision, considering that the ball travels for 5 seconds until it hits the wall?

A. $E = 780m$ J   B. $E = 500m$ J   C. $E = 1300m$ J

Question 5.1

A lift with a mass of 800 kg can carry up to 700 kg of passengers. Calculate the total energy needed for an electric motor is used to raise the elevator with a full load from the ground floor to the third floor which is 7 m higher. You may assume g =10 m/s².

A. 105 J   C. 105 kJ   E. 105 MJ

B. 210 J   D. 210 kJ

Question 5.2

Calculate the power of motor required to do this in 30 s.

A. 700 W   C. 7 kW   E. 70 kW

B. 3500 W   D. 35 kW

Question 5.3

Find the average kinetic energy of the lift and passengers during the ascent.

A. 41 J   C. 67 J   E. 105 J

B. 52 J   D. 82 J

Question 5.4

The elevator takes 10 seconds to accelerate and 10 seconds to decelerate using the same magnitude of constant force in each case. Between acceleration and deceleration, no force is used. Calculate the average speed which the lift attains.

A. 0.04 ms⁻¹   C. 0.35 ms⁻¹   E. 2.3 ms⁻¹

B. 0.14 ms⁻¹   D. 1.3 ms⁻¹

Question 6.1

If the friction coefficient between m_1 and m_2 is μ_1, and between m_2 and the inclined plane is μ_2, (where α = the angle of inclination) determine the acceleration of m_2.

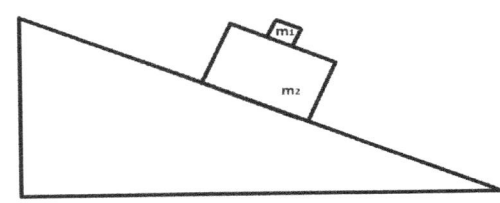

A. $g(1 - \mu_2\cos\alpha)$

B. $g(\mu_2\sin\alpha - \mu_1\cos\alpha)$

C. $mg(\sin\alpha - \mu_2\cos\alpha)$

D. $g(\sin\alpha - \mu_1\cos\alpha)$

E. $g(\sin\alpha - \mu_2\cos\alpha)$

Question 6.2

Determine the acceleration of m_1 with respect to the plane.

A. $g(\sin\alpha - \mu_1\cos\alpha)$

B. $mg(\sin\alpha - \mu_1\cos\alpha)$

C. $g\sin\alpha(\mu_1 - \mu_2)$

D. $g(\sin\alpha - \cos\alpha)$

E. $g(\mu_2\sin\alpha - \mu_1\cos\alpha)$

Question 6.3

Determine the acceleration of m_1 with respect to m_2.

A. $g\cos\alpha(\mu_1 + \mu_2)$

B. $g\cos\alpha(\mu_2 - \mu_1)$

C. $g\sin\alpha(\mu_1 - \mu_2)$

D. $mg\cos\alpha(\mu_1 - \mu_2)$

E. $mg\cos\alpha(\mu_1 + \mu_2)$

Question 6.4

Which of the following would give the coefficient of friction, μ_2?

A. $\tan(\alpha)$

B. $\cos(\alpha)$

C. $\sin(\alpha)$

D. mg

E. $mg\tan(\alpha)$

Question 7.1:

The graph below represents the velocity as a function of time of a car (with mass 1000 kg) moving in a straight line. How far has the car travelled after 30 min?

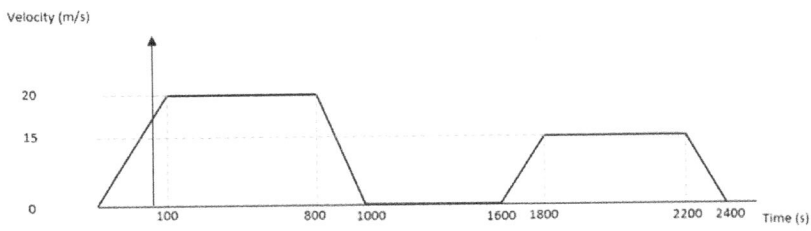

A. 4900 m

B. 9000 m

C. 20000 m

D. 18500 m

E. 5900 m

Question 7.2:

Compute the following calculation$\int_{1600}^{1800}$ v dt. Select the option that represents the correct corresponding value and its physical meaning.

A. Acceleration: 1500 ms^{-2}

B. Acceleration: 2300 ms^{-2}

C. Velocity: 1500 ms^{-1}

D. Distance: 1000 m.

E. Distance: 1500 m.

Question 7.3:

Calculate the force exerted on the car at t = 900s.

A. 110 N

B. 100 N

C. 75 N

D. 500 N

E. 765 N

Question 7.4:

What is the power delivered by the car's engine in the interval of 0 – 100 s?

A. 2 kW

B. 900 W

C. 1.5 kW

D. 2.5 kW

E. 1.05 kW

Question 8.1:

The following diagram represents the horizontal component of a force acting on a particle as a function of its position. If the particle is at rest at x = 0, what is its position (along the x-axis) when its velocity is maximum?

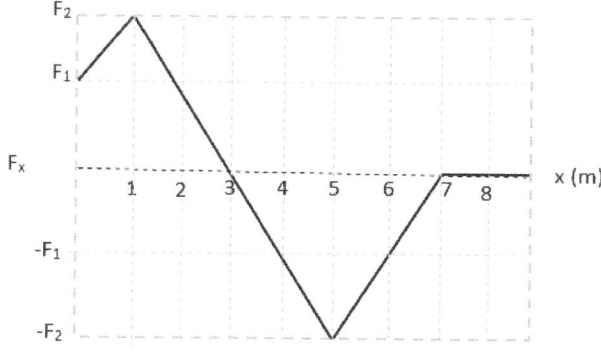

A. 1 m

B. 5 m

C. 3 m

D. 4 m

E. 8 m

Question 8.2: What is the particle's position (along the x-axis) when its kinetic energy is maximum?

A. 1 m

B. 5 m

C. 7 m

D. 3 m

E. 8 m

Question 8.3: What is the particle' position (along the x-axis) when its velocity is zero?

A. 6 m
B. 1 m

C. 7 m
D. 3 m

E. 8 m

Question 8.4: After x = 6 metres, which of the following statements is true?

A. The velocity vector is pointing in the same direction as it is in x = 3m.
B. The velocity vector is pointing in the same direction as it is in x = 2m.
C. The acceleration vector is pointing in the same direction as it is in x = 2m.
D. The acceleration vector is point in the opposite direction as it is in x = 2m.

Question 9.1: A little girl throws vertically a ball, of mass 0.5 kg, into the air at t =0. Which of the following graphs best describe the velocity of the ball as a function of time? Disregard the influence of air in the movement of the ball.

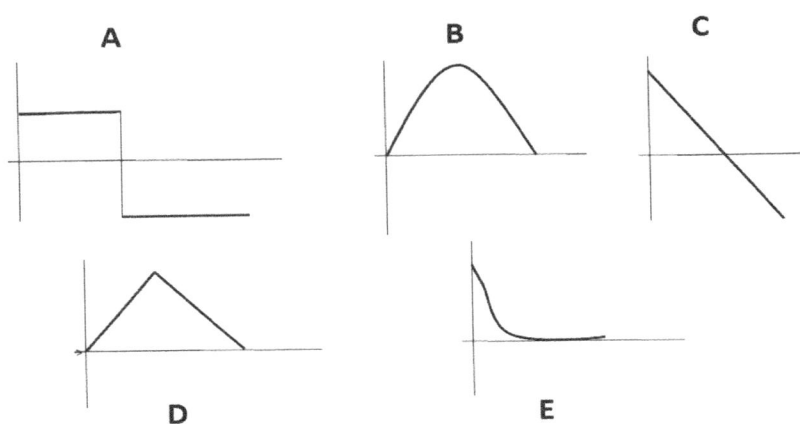

Question 9.2: Given that the ball leaves the girl's hand at 15 m/s, what is the approximate highest point from the ground the ball can reach? Disregard the influence of air.

A. 6 m
B. 12 m

C. 25 m
D. 3.5m

E. 8 m

Question 9.3: Approximately, how long does the ball take to reach the highest point of its trajectory?

A. 1.5 s
B. 10 s

C. 5 s
D. 4 s

E. 9.5 s

Question 9.4: What would the potential and kinetic energy of the ball be, respectively, 5.0 m from the ground?

A. 31.75 J and 24.5 J
B. 25.5 J and 37.75 J

C. 24.5 J and 31.75 J
D. 30 J and 26 J

E. 11 J and 27 J

Question 10.1: A horizontal force is applied to the ramp (Object 1), of mass 10 kg, as shown in the figure. There is no friction between the ramp and the ground. The coefficient of friction between the ramp and Object 2, also of mass 10 kg, is $\mu = 0.2$. What option correctly demonstrates the forces applied on Object 2?

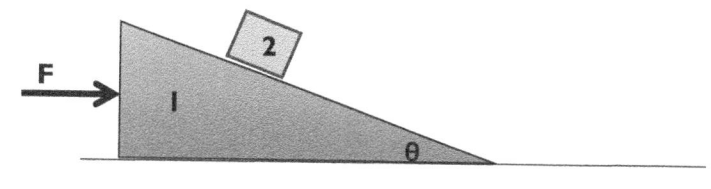

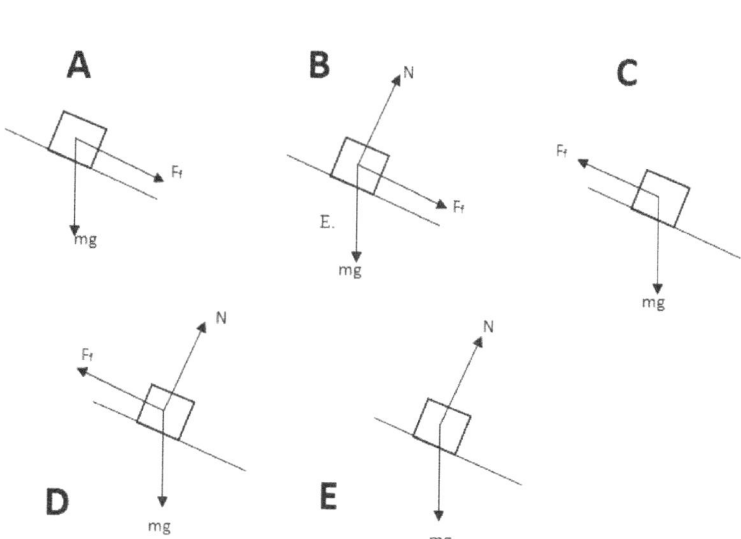

Question 10.2: Express the weight of Object 2 as a function of the angle θ and the coefficient of friction μ.

A. $mg = N\mu(\sin\theta - \cos\theta)$ C. $mg = N(\mu\sin\theta - \cos\theta)$ E. $mg = N(\cos\theta - \mu\sin\theta)$
B. $mg = N\mu(\cos\theta + \sin\theta)$ D. $mg = \mu(\sin\theta + N\cos\theta)$

Question 10.3: Express the acceleration of Object 2, in relation to the Object 1, as a function of the angle θ and the coefficient of friction μ.

A. $a = N(\sin\theta - \mu\cos\theta)$ C. $a = \dfrac{N}{m_2}(\cos\theta + \mu\cos\theta)$ E. $a = \dfrac{N}{m_2}(\cos\theta + \mu\sin\theta)$
B. $a = \dfrac{N}{m_2}(\sin\theta - \mu\cos\theta)$ D. $a = \dfrac{N}{m_2}(\sin\theta + \mu\cos\theta)$

Question 10.4: What is the maximum value of the force F, in Newtons, that can be applied without provoking the movement of Object 2 upwards along the ramp? Consider $\sin\theta = 0.6$ and $\cos\theta = 0.8$.

A. 219 N   C. 200N   E. 149 N
B. 350 N   D. 155 N

ANSWERS

ANSWER KEY

Q	A	Q	A	Q	A	Q	A	Q	A	Q	A	Q	A
1	F	31	C	61	D	91	C	121	B	151	D	181	C
2	A	32	G	62	F	92	A	122	A	152	D	182	C
3	D	33	D	63	B	93	C	123	F	153	B	183	E
4	E	34	D	64	A	94	B	124	D	154	E	184	C
5	G	35	E	65	C	95	A	125	A	155	E	185	C
6	C	36	B	66	C	96	A	126	B	156	B	186	A
7	D	37	A	67	G	97	D	127	A	157	C	187	D
8	E	38	E	68	C	98	C	128	F	158	A	188	C
9	D	39	D	69	B	99	C	129	D	159	C	189	B
10	D	40	F	70	B	100	C	130	A	160	B	190	A
11	F	41	E	71	E	101	B	131	D	161	B	191	C
12	B	42	A	72	C	102	C	132	D	162	B	192	C
13	C	43	B	73	E	103	C	133	F	163	C	193	E
14	G	44	C	74	E	104	C	134	B	164	C	194	E
15	ED	45	D	75	D	105	E	135	C	165	A	195	A
16	E	46	F	76	E	106	A	136	B	166	C	196	C
17	A	47	E	77	E	107	C	137	C	167	D	197	C
18	E	48	C	78	C	108	E	138	A	168	C	198	E
19	G	49	C	79	E	109	E	139	A	169	D	199	A
20	E	50	B	80	D	110	C	140	C	170	A	200	D
21	H	51	D	81	E	111	E	141	B	171	C	201	C
22	E	52	F	82	C	112	E	142	D	172	B	202	D
23	FG	53	G	83	C	113	E	143	C	173	B	203	A
24	D	54	D	84	B	114	B	144	A	174	A	204	D
25	E	55	D	85	B	115	C	145	C	175	C	205	B
26	D	56	A	86	D	116	B	146	C	176	C	206	E
27	G	57	E	87	B	117	B	147	C	177	B	207	A
28	F	58	C	88	E	118	C	148	B	178	A	208	D
29	B	59	D	89	C	119	D	149	D	179	C	209	C
30	A	60	C	90	A	120	C	150	E	180	E	210	E

SECTION 1: WORKED ANSWERS

Question 1: F
Statement F, that the amplitude of a wave determines its mass, is false. Waves are continuous, not particulate, and do not have mass.

Question 2: A
We are given that displacement s = 30 m, initial speed u = 0 ms^{-1} and acceleration a = 5.4 ms^{-2}. We are asked to find the final speed v, and the time taken t.

We require the SUVAT equation without time to find final velocity, which is $v^2 = u^2 + 2as$.

$$v^2 = u^2 + 2as \Rightarrow v^2 = 0^2 + 2 \times 5.4 \times 30$$

$$\Rightarrow v^2 = 324 \therefore v = 18 \, ms^{-1}$$

Next, to find the time, we can use:

$$s = ut + \frac{1}{2}at^2 \Rightarrow 30 = 0 \times t + \frac{1}{2} \times 5.4 \times t^2$$

$$\Rightarrow t^2 = \frac{30}{2.7} = 3.3 \, s$$

Question 3: D
The time period is calculated using the fact that the canoe rises and falls 7 times within 49 seconds. Therefore, T = 49/7 = 7 seconds.

To find the wavelength, we need to use the $v = f\lambda$ equation. Frequency is the reciprocal of T, so it is given by 1/T = 1/7 s^{-1}. Therefore:

$$\lambda = \frac{v}{f} = \frac{5}{1/7} = 35 \text{ m}.$$

Question 4: E
This is a straightforward question, as there is an equation which directly links all the provided quantities. The only challenging aspect is the numbers being more difficult in the calculations.

$$\text{Power} = \frac{\text{Force x Distance}}{\text{Time}} = \frac{375 \text{ N x } 1.3 \text{ m}}{5 \text{ s}} = 75 \times 1.3 = 97.5 \text{ W}$$

Question 5: G

As there is constant acceleration, we need to use a SUVAT equation. We have been given the initial velocity, time taken and the acceleration.

Therefore, we need to use $v = u + at$ to find the final velocity.

$$v = u + at = 0 + 5.6 \times 8 = 44.8 \, ms^{-1}$$

And, to find distance, we need to use:

$$s = ut + \frac{at^2}{2} = 0 + 5.6 \times \frac{8^2}{2} = 179.2 \, m$$

Question 6: C

The sky diver leaves the plane and will accelerate until the air resistance equals their weight – this is their terminal velocity. The sky diver will accelerate under the force of gravity. If the air resistance force exceeded the force of gravity the sky diver would accelerate away from the ground, and if it were less than the force of gravity they would continue to accelerate towards the ground.

Question 7: D

We have been given that s = 20 m, u = 0 ms⁻¹ and a = 10 ms⁻². To find the momentum before impact, we need to know the apple's velocity before impact. Therefore, we wish to find the final velocity and the equation required is:

$$v^2 = u^2 + 2as \Rightarrow v^2 = 0 + 2 \times 10 \times 20$$

$$\Rightarrow v^2 = 400 \therefore v = 20 \, ms^{-1}.$$

Therefore, the momentum is given by:

$$p = mv = 20 \times 0.1 = 2 \, kgms^{-1}.$$

Question 8: E

Electromagnetic waves have varying wavelengths and frequencies, and, as their energy is proportional to their frequency, have varying energies. Therefore, only statements 1, 2 and 5 are correct.

Question 9: D

The total resistance = R + r = 0.8 + 1 = 1.8 Ω. Therefore, the current can be calculated as follows:

$$I = \frac{EMF}{Total \, Resistance} = \frac{36}{1.8} = 20 \, A.$$

Question 10: D

Use Newton's second law and remember to work in SI units:

$$F = \frac{\Delta p}{\Delta t} = \frac{\Delta(mv)}{\Delta t} = m\frac{\Delta v}{\Delta t}.$$

The change in velocity is given in the question – it is 100 ms⁻¹ as the dart is brought to rest. Therefore, the average force is given by:

$$\Rightarrow F = 20 \times 10^{-3} \times \frac{100}{10 \times 10^{-3}} = 200 \, N.$$

Question 11: F

As the Professor is lifting the bag against gravity, work is being done. This work done is given by:

$$\text{Work Done} = F \times d = \text{Bag's Weight} \times \text{Distance} = 50 \times 10 \times 0.7 = 350 \text{ N}.$$

The average power, therefore, is given by:

$$\text{Power} = \frac{\text{Work}}{\text{Time}} = \frac{350}{3} = 116.7 \text{ W}.$$

This gives a value of 117 W to 3 significant figures, so the correct answer is F.

Question 12: B

The engine is supplying the driving force – therefore, to calculate the current, we are only concerned with the driving force. To find the power supplied by the engine, we need to use:

$$P = Fv = 300 \text{ N} \times 30 \text{ ms}^{-1} = 9000 \text{ W}.$$

To calculate the current, we need to use:

$$P = IV \Rightarrow I = \frac{P}{V} = \frac{9000}{200} = 45 \text{ A}.$$

Question 13: C

Work is defined as the magnitude of the force multiplied by the displacement in the direction of the force. Therefore, statement 3 is correct. Statement 1, therefore, is not correct as it suggests you divide the force by the displacement. For statement 2, we need to work out the units of work. Using the equation, we can see that:

$$W = F \times d \Rightarrow W = ma \times d$$

By considering the units, we can clearly see that:

$$[W] = kg \times ms^{-2} \times m = kgm^2s^{-2}.$$

Therefore, statement 2 is incorrect. Thus, only statement 3 is correct.

Question 14: F

Joules are the unit of energy – using the equation W = F × d, we can obtain that 1 Joule = 1 N x 1 m.

Pa is the unit of Pressure - therefore, by using the following equation, we obtain that:

$$P = \frac{Force}{Area} = \frac{N}{m^2}.$$

$$\therefore Pa \times m^3 = \frac{Nm^3}{m^2} = Nm.$$

Therefore, statement 2 is indeed true. Statement 1 is also true as this is the defining equation for kinetic energy.

The kinetic energy that a body carries is the total energy it has due to its movement. Therefore, if we were to bring the object to rest, it would lose exactly its kinetic energy – as this is the energy 'locked-up' in its motion. Therefore, all 3 statements are true.

Question 15: E

Radiation can indeed be in the form of waves or particles, and may be ionising depending on its energy. Gamma radiation does have very high energy, followed by beta radiation and then alpha radiation. Statement E, however, is not true as X-rays are waves and, therefore, are not an example of particle radiation.

Question 16: E

Statement 1 is false as the half-life depends on the physical properties of the atom – the atom type and isotope clearly affect the physical properties of the atom. Statement 2 is the correct definition of half-life.

Statement 3 is also correct: half-life in exponential decay will always have the same duration, independent of the quantity of the matter in question – this is a defining feature of exponential decay. In non-exponential decay, half-life is dependent on the quantity of matter in question.

Question 17: A

In contrast to nuclear fission, where neutrons are shot at unstable atoms, nuclear fusion is based on the high speed, high-temperature collision of molecules, most commonly hydrogen, to form a new, heavier atoms while releasing energy. Therefore, all statements are true except statement A, which is the case for nuclear fission.

Question 18: E

Nuclear fission releases a significant amount of energy, which is the basis of many nuclear weapons. Shooting neutrons at unstable atoms destabilises the nuclei which in turn leads to a chain reaction and fission. Nuclear fission can lead to the release of ionizing gamma radiation. Therefore, all of the statements are true.

Question 19: G

As the two resistors are connected in a series circuit, the current is the same at all points – therefore, statement 1 is true. Due to Ohm's Law, the potential difference across the two resistors must also be the same – as the current and resistance of each resistor are equal. Therefore, statement 2 is also true. Ohm's Law can also be applied to the two resistors together, so statement 3 is also correct.

Question 20: E

To find the circumference, the radius of the orbit needs to be evaluated first. To determine the radius, we have to convert the time into seconds and then use the following equation:

$$Speed = \frac{Distance}{Time} \Rightarrow Distance = Speed \times Time.$$

$$\therefore Distance = 3 \times 10^8 \times (60 \times 8) = 1440 \times 10^8 \approx 1.5 \times 10^{11} \, m$$

As the answer only requires the power of ten, we can take this as 10^{11} metres. To find the circumference, we need to use:

$$C = 2\pi r = 2\pi \times (10^{11}) \approx 10^{12} \text{ m.}$$

Question 21: H
Speed is a scalar quantity whilst velocity is a vector, which means it describes both magnitude and direction. Speed describes the distance a moving object covers over time (i.e. speed = distance/time), whereas velocity describes the rate of change of the displacement of an object (i.e. velocity = displacement/time). The SI unit for speed is meters per second (ms^{-1}), while ms^{-2} is the standard unit of acceleration.

Question 22: E
Ohm's Law only applies to conductors as insulators cannot carry a current. The law can be mathematically expressed as $V = IR$. This suggest that potential difference is directly proportional to current, which corresponds to statement E.

Question 23: F
Any object at rest is not accelerating and therefore has no resultant force acting upon it. Therefore, statement 1 is indeed true.

Strictly speaking, Newton's second law is given by:

$$F = \frac{dp}{dt} = \text{Rate of change of momentum.}$$

This can be rearranged as follows:

$$F = \frac{dp}{dt} = \frac{d(mv)}{dt} = m\frac{dv}{dt} + v\frac{dm}{dt}.$$

In the case of constant mass, this reduces to:

$$F = m\frac{dv}{dt} = ma.$$

Therefore, this is not the most general form of Newton's Second Law and is only applicable in the cases of constant mass. Statement 2, therefore, is false.

Question 24: D
Equation 1 is correct, as charge equals current multiplied by time. By substituting in Ohm's Law, which is $I = \frac{V}{R}$, equation 1 is obtained. Equally, the equation $I = \frac{P}{V}$ may be substituted to reproduce equation 2. Equation 3, however, is incorrect as charge is simply equal to the numerator of the equation – the division by resistance is incorrect.

Question 25: E
The total weight of the elevator and the people within it is given by:

$$W = mg = 10 \times (1600 + 200) = 18,000 \, \text{N}.$$

The total downwards force on the elevator is the sum of the frictional force and the weight. The motor force has to act in the upwards direction. Applying Newton's second law of motion on the car gives:

$$F_M = \text{Motor Force} - [\text{Frictional Force} + \text{Weight}]$$

$$F_M = M - 4,000 - 18,000 = M - 22,000$$

To get an acceleration of 1 ms^{-2} upwards, this must be equal to:

$$F_M = ma = 1800 \times 1 = 1800 \, N$$

$$\Rightarrow 1800 \, N = M - 22{,}000 \quad \therefore M = 23{,}800 \, N.$$

Question 26: D

The total distance travelled is the sum of the distance travelled during acceleration phase and the distance travelled during the braking phase. The distance during <u>acceleration phase</u> is given by:

$$s = ut + \frac{at^2}{2} = 0 + \frac{5 \times 10^2}{2} = 250 \text{ m.}$$

To find the distance during the second phase, we need to find the final velocity:

$$v = u + at = 0 + 5 \times 10 = 50 \text{ ms}^{-1}.$$

Now, we can use $a = \frac{v-u}{t}$ to calculate the deceleration:

$$a = \frac{0 - 50}{20} = -2.5 \text{ ms}^{-2}.$$

Therefore, the distance travelled during the <u>deceleration phase</u> is given by:

$$s = ut + \frac{at^2}{2} = 50 \times 20 + \frac{-2.5 \times 20^2}{2} = 1000 - \frac{2.5 \times 400}{2}$$

$$\Rightarrow s = 1000 - 500 = 500 \text{ m.}$$

Therefore, the total distance is the sum of these two distances, which gives:

$$\text{Total Distance} = 250 + 500 = 750 \text{ m.}$$

Question 27: G

It is not possible to calculate the power of the heater as we do not know the current that flows through it or its internal resistance. The 8 ohms refers to the external copper wire and not the heater. Whilst it is important that you know how to use equations like $P = IV$, it's more important that you know when you *can't* use them!

Question 28: F

This question has a lot of information but there is no mention of the duration of each pulse. The quantity of time for which the electrons are accelerated is necessary to calculate power. Similarly, you cannot calculate power by using $P = IV$ as you do not know how many electrons are accelerated through the potential difference per unit time. Thus, more information is required to calculate the power.

Question 29: B
When an object is in equilibrium with its surroundings, it radiates and absorbs energy at the same rate and so its temperature remains constant. In other words, there is no *net* energy transfer. Therefore, statement A is false.

Radiation is slower than conduction and convection. Therefore, statements C and D are both false.

Question 30: A
The work done by the force is given by:

$$\text{Work Done} = \text{Force} \times \text{Distance} = 12\,\text{N} \times 3\,\text{m} = 36\,\text{J}.$$

Since the surface is frictionless, we can further state that:

$$\text{Work Done} = \text{Kinetic Energy} \Rightarrow E_k = \frac{mv^2}{2} = \frac{6v^2}{2}.$$

$$\therefore 36 = 3v^2 \Rightarrow v = \sqrt{12} = \sqrt{4}\sqrt{3} = 2\sqrt{3}\,\text{ms}^{-1}.$$

Question 31: C
The total energy supplied to the water is the product of the change in the temperature, the mass of the water and its specific heat capacity, which is 4000 J in this case. Therefore:

$$Q = mc\,\Delta T = \text{Change in temperature} \times \text{Mass of water} \times 4{,}000\,\text{J}.$$

$$\Rightarrow Q = 40 \times 1.5 \times 4{,}000 = 240{,}000\,\text{J}.$$

The power of the heater can also be evaluated with the quantities given, which allows us to determine the resistance using the $P = \frac{V^2}{R}$ relation:

$$P = \frac{\text{Work Done}}{\text{time}} = \frac{240{,}000}{50 \times 60} = \frac{240{,}000}{3{,}000} = 80\,\text{W}.$$

$$P = IV = \frac{V^2}{R} \Rightarrow R = \frac{V^2}{P} = \frac{100^2}{80} = \frac{10{,}000}{80} = 125\,\Omega.$$

Question 32: G
Nuclear power plants utilise the large amount of energy released during atomic fission, therefore statement I is indeed true. Splitting an atom into two or more parts will, by definition, produce molecules of different sizes than the original atom – as these are the constituents of the parent particle. The free neutrons and photons produced by the splitting of atoms form the basis of the energy release.

Question 33: D

Gravitational potential energy is just an extension of the equation work done equals force multiplied by distance (force is the weight of the object, *mg*, and distance is the height, *h*). Therefore, statements 1 and 2 are both correct. The reservoir in statement 3 would have a potential energy of 10^{10} Joules, which is equivalent to 10 Giga Joules, as:

$$E_p = 10^6 \text{ kg} \times 10 \text{ N} \times 10^3 \text{ m} = 10 \text{ GJ}.$$

Question 34: D

Statement 1 is the common formulation of Newton's third law. Statement 2 presents a consequence of the application of Newton's third law.

Statement 3 is false: rockets can still accelerate because the products of burning fuel are ejected in the opposite direction from which the rocket needs to accelerate. Therefore, the fuel experiences the equal and opposite force to the rocket.

Question 35: E

Positively charged objects have lost electrons, as electrons are negatively charged particles. Therefore, statement 1 is false. For the second statement, we can recall the following equation:

$$\text{Charge} = \text{Current} \times \text{Time} = \frac{\text{Voltage}}{\text{Resistance}} \times \text{Time}.$$

Therefore, we can evaluate the amount of charge when provided with the period of time, voltage and resistance. Lastly, objects can become charged by friction as electrons are transferred from one object to the other.

Question 36: B

Each body of mass exerts a gravitational force on another body with mass. This is true for all planets, as well as all bodies with mass in the universe. Gravitational force is indeed dependent on the mass of both objects, which is evident from the gravitational force equation.

Satellites stay in orbit due to centripetal force that acts tangentially to gravity (not because of the thrust from their engines). Two objects will only land at the same time if they also have the same shape as otherwise air resistance would result in different terminal velocities. Note that this is not relevant in space as there is no air resistance.

Question 37: A

Metals conduct electrical charge easily and provide little resistance to the flow of electrons. Charge can also flow in several directions, although will typically have a general direction of motion in a circuit. Lastly, all conductors have an internal resistance and therefore provide *some* resistance to electrical charge.

Question 38: E

First, we can calculate the rate of petrol consumption:

$$\frac{\text{Speed}}{\text{Consumption}} = \frac{60 \text{ miles/hour}}{30 \text{ miles/gallon}} = 2 \text{ gallons/hour}.$$

Therefore, the amount of energy required in one hour is: 2 gallons $= 2 \times 9 \times 10^8 = 18 \times 10^8$ J. As 1 hour $= 60 \times 60 = 3600$ seconds, we can use the following equation to determine the power:

$$\text{Power} = \frac{\text{Energy}}{\text{Time}} = \frac{18 \times 10^8}{3600} \Rightarrow P = \frac{18}{36} \times 10^6 = 5 \times 10^5 \text{ W}.$$

As only 20% of the power is delivered to the wheels, we need to find 20% of the total power.

$$\therefore P_{wheels} = 5 \times 10^5 \times 0.2 = 10^5 \text{W} = 100 \text{ kW}.$$

Question 39: F

Beta radiation is stopped by a few millimetres of aluminium, but not by paper. Therefore, statement 2 is correct. In β^- radiation, a neutron changes into a proton plus an emitted electron. This means the atomic mass number remains unchanged. Therefore, statement 1 is also correct. Finally, statement 3 is also correct – as beta particles are charged, they are deflected within electric fields. As they are also in motion, they experience a magnetic force and so also deflect within a magnetic field.

Question 40: F

Firstly, we can calculate the mass of the car using the following relation: Mass $= \frac{\text{Weight}}{g} = \frac{15,000}{10} = 1,500$ kg.

The average braking force is the product of the mass of the car and its deceleration. To calculate the deceleration of the car, we can use the following equation, where v = 0 ms^{-1}, u = 15 ms^{-1} and t = 10 × 10^{-3} s:

$$v = u + at \Rightarrow a = \frac{v-u}{t} = \frac{0-15}{0.01} = 1500 \text{ ms}^{-2}.$$

Therefore, the average braking force is given by:

F = ma = 1500 × 1500 = 2 250 000 N.

Question 41: E

Electrical insulators offer high resistance to the flow of charge. Insulators are usually non-metals, as metals conduct charge very easily due to their sea of delocalised electrons. Insulators can indeed be charged by friction as charge does not flow easily within the material – therefore, it cannot cancel the charge out immediately like a conductor.

Question 42: A

The car accelerates for the first 10 seconds at a constant rate, which is given by the gradient of the first linear region, and then decelerates after t = 30 seconds, which is given by the gradient of the second linear region. It does not reverse, as the velocity is not negative. Therefore, only statement 1 is incorrect.

Question 43: B

The distance travelled by the car is represented by the area under the curve (integral of velocity) which is given by the area of two triangles and a rectangle:

$$\text{Area} = \left(\frac{1}{2} \times 10 \times 10\right) + (20 \times 10) + \left(\frac{1}{2} \times 10 \times 10\right)$$

$$\Rightarrow \text{Area} = 50 + 200 + 50 = 300 \text{ m.}$$

Question 44: C

We can use Newton's second to determine the change in velocity. We know that F = 10,000 N, mass = 1,000 kg and change in time is 5 seconds, so these can be substituted.

$$F = \frac{\Delta p}{\Delta t} = m\frac{\Delta v}{\Delta t} \Rightarrow \Delta v = \frac{F \times \Delta t}{m} = \frac{10,000 \times 5}{1,000} = 50 \, ms^{-1}.$$

Question 45: D

This question tests both your ability to convert unusual units into SI units and to select the relevant values (for example, the crane's mass is not important here). Firstly, the conversions are: 0.01 tonnes = 10 kg, 100 cm = 1 metre and 5,000 milliseconds = 5 seconds.

To calculate power, we need to use the following equation:

$$\text{Power} = \frac{\text{Work Done}}{\text{Time}} = \frac{\text{Force x Distance}}{\text{Time}}$$

The force that the crane is working against is the weight of the car. This is easily evaluated:

$$\text{Weight of the car} = 10 \times g = 10 \times 10 = 100N \Rightarrow \text{Power} = \frac{100 \times 1}{5} = 20 \text{ W.}$$

Question 46: F

The total resistance, denoted R_T, in a parallel circuit is given by:

$$\frac{1}{R_T} = \frac{1}{R_1} + \frac{1}{R_2} + \cdots$$

Thus, we can substitute in the values provided for the two resistors:

$$\Rightarrow \frac{1}{R_T} = \frac{1}{1} + \frac{1}{2} = \frac{3}{2} \quad \therefore R_T = \frac{2}{3}\Omega$$

Using Ohm's Law, we can obtain the current:

$$I = \frac{20 \text{ V}}{\frac{2}{3}\Omega} = 20 \times \frac{3}{2} = 30 \text{ A.}$$

Question 47: E

Water has a greater refractive index than air; therefore, the speed of light decreases when it enters water and increases when it leaves water. The direction of light also changes when light enters or leaves water. This phenomenon is known as refraction and is governed by Snell's Law.

Question 48: C
The voltage in a parallel circuit is equal across each branch. Therefore, the voltage of branch A equals that of branch B. The resistance of the two branches are obtained by multiplying the number of resistors in each branch by the individual resistance:

$R_A = 6 \times 5 = 30\ \Omega, \quad R_B = 10 \times 2 = 20\ \Omega.$

Using Ohm's Law, the current across each branch is easily evaluated:

$V = IR \Rightarrow I = \dfrac{V}{R} \quad \therefore \quad I_A = \dfrac{60}{30} = 2\ \text{A}, \ I_B = \dfrac{60}{20} = 3\ \text{A}.$

Question 49: C
This is a very straightforward question, but it is slightly challenging due to the awkward units provided. Ensure you are able to work comfortably with prefixes of 10^9 and 10^{-9}, and convert between the relevant units correctly. The conversions required are:

50,000,000,000 nanowatts = 50 W, 0.000000004 giga-amps = 4 A.

Using the $P = IV$ relation, it is straightforward to determine the voltage of the circuit:

$V = \dfrac{P}{I} = \dfrac{50}{4} = 12.5\ \text{V} = 0.0125\ \text{kV}.$

Question 50: B
Radioactive decay of a single atom is highly random and unpredictable - the behaviour of a large group of atoms can be successfully modelled using a half-life, but it is not possible to predict when a single atom will decay. Therefore, statement A is false.

Only gamma decay releases gamma rays and few types of decay release X-rays. Therefore, both statements C and D are incorrect. The electrical charge of an atom's nucleus decreases after alpha decay as two protons are lost. Therefore, statement E is also incorrect.

Question 51: D
There is more than one way to approach this question. You can calculate current and use that to find the total resistance, or you can use algebra by denoting the unknown resistance and using the relevant power equation. The former method is as follows:

$P = IV \Rightarrow I = \dfrac{P}{V} = \dfrac{60}{15} = 4\ \text{A}.$

This is the current through the circuit, so we can utilise Ohm's Law to find the total resistance of the circuit as follows:

$R = \dfrac{V}{I} = \dfrac{15}{4} = 3.75\ \Omega.$

So each resistor has a resistance of $\dfrac{3.75}{3} = 1.25\ \Omega.$

If two more resistors are added, the overall resistance $R_T = 1.25 \times 5 = 6.25\ \Omega.$

Question 52: F

To calculate the useful work done, and hence the efficiency, we require the useful output over the total energy inputted. In this case, we are not provided with the resistive forces on the tractor, whether it is stationary or moving at the end point, and if there is any change in vertical height.

Question 53: G

Electromagnetic induction is defined by statements 1 and 2. An electrical current is generated when a coil moves in a magnetic field. In essence, there needs to be relative motion between a conductor, such as the wire, and a magnetic field, which is satisfied in all three cases.

Question 54: D

An ammeter will always give the same reading in a series circuit, in which the current is the same at all points, but not in a parallel circuit where current splits at each branch in accordance with Ohm's Law. Therefore, the current does indeed depend on the resistance of that branch. It is also conserved – so the sum of the currents in the branches is equal to the total current within the circuit.

Question 55: D

Electrons move in the opposite direction to the conventional current (they move from negative to positive) as they are attracted to the positive terminal due to their negative charge. Therefore, statement 3 is incorrect.

Question 56: A

For a fixed resistor, the current is directly proportional to the potential difference. For a filament lamp, the metal filament becomes hotter as the current increases. This causes the metal atoms to vibrate and move more, resulting in more collisions with the flow of electrons. This makes it harder for the electrons to move through the lamp and results in increased resistance. Therefore, the graph's gradient decreases as current increases.

Question 57: E

Vector quantities consist of both direction and magnitude, and can be added by taking account of the direction in the sum. They can also be subtracted by reversing their direction. Lastly, displacement is indeed a vector.

Question 58: C

The gravity on the moon is 6 times weaker than that on Earth. Therefore:

$g_{Earth} = 10 \text{ ms}^{-2} \Rightarrow g_{moon} = \frac{10}{6} = \frac{5}{3} \text{ ms}^{-2}.$

Since Weight = Mass × Gravity, the mass of the rock is given by:

$m_{rock} = \frac{250}{\frac{5}{3}} = \frac{750}{5} = 150 \text{ kg}.$

Therefore, the density of the rock is given by the following:

$\text{Density} = \frac{\text{mass}}{\text{volume}} = \frac{150}{250} = 0.6 \text{ kg/cm}^3.$

Question 59: D

An alpha particle consists of a helium nucleus. Thus, alpha decay causes the mass number to decrease by 4 and the atomic number to decrease by 2. Five iterations of this would decrease the mass number by 20 and the atomic number by 10. This would result in a mass number of $225 - (4 \times 5) = 205$ and an atomic number of $78 - (2 \times 5) = 68$.

Question 60: C

The circuit is connected to the primary coil in the transformer. Therefore, to find the potential difference in the primary coil, we can use Ohm's law:

$$V_{primary} = IR = 20 \times 10 = 200 \text{ V}.$$

Now the equation provided can to be utilised to determine the potential difference exiting the transformer:

$$\frac{N1}{N2} = \frac{V1}{V2} \Rightarrow \frac{5}{10} = \frac{200}{V2} \Rightarrow V_{secondary} = \frac{2,000}{5} = 400 \text{ V}.$$

Question 61: D

For objects in free fall that have reached terminal velocity, the weight is exactly equal and opposite to the air resistance. To find the mass of the sphere, the weight needs to be determined. This can be done using the work done by the resistive forces as follows:

$$\text{Work Done} = \text{Force} \times \text{Distance} \Rightarrow \text{Force} = \frac{10,000 \text{ J}}{100 \text{ m}} = 100 \text{ N}.$$

Therefore, the sphere's weight = 100 N, as it balances this resistive force. Since $g = 10 \text{ ms}^{-2}$, the sphere's mass is simply:

$$\text{Mass} = \frac{\text{Weight}}{g} = \frac{100}{10} = 10 \text{ kg}.$$

Question 62: F

The wavelength of ultraviolet waves is longer than that of x-rays. Wavelength is inversely proportional to frequency. Most electromagnetic waves are not stopped with aluminium (and require thick lead to stop them), and they travel at the speed of light. Humans can only see a very small part of the spectrum. Therefore, all of the statements are incorrect.

Question 63: B

If an object moves towards the sensor, the wavelength will appear to decrease and the frequency increase. The faster the rate of approach, the faster the increase in frequency and decrease in wavelength. Alternatively, when the object moves away, we expect the frequency to decrease at that rate.

Question 64: A

The bullet has an initial velocity of 1000 m/s, and is brought to rest in a time on 0.1 s. Therefore, the deceleration of the bullet, due the wall, is given by:

$$\text{Deceleration} = \frac{\text{Change in Velocity}}{\text{Time}} = \frac{v - u}{t} = \frac{1,000}{0.1} = 10,000 \text{ ms}^{-2}.$$

Using Newton's second law, the braking force provided by the brick wall can be evaluated:

$$\text{Braking Force} = \text{Mass} \times \text{Acceleration} = 10,000 \times 0.005 = 50 \text{ N}.$$

Question 65: C

Polonium has undergone alpha decay, as its mass number and atomic number have both changed due to the decay. Only alpha decays have this property. Thus, Y is a helium nucleus and contains 2 protons and 2 neutrons.

Therefore, as there are 6×10^{23} particles in a single mole, each with two protons, 10 moles of Y contain:

Number of protons $= 2 \times 10 \times (6 \times 10^{23})$ protons $= 120 \times 10^{23} = 1.2 \times 10^{25}$ protons.

Question 66: C

The rod's activity is less than 1,000 Bq after 300 days. To calculate the longest possible half-life, we must assume that the activity is just below 1,000 Bq after 300 days. Thus, the half-life has decreased activity from 16,000 Bq to 1,000 Bq during this period of time.

After one half-life, we would expect the activity to decrease to 8,000 Bq. After two half-lives, it should be 4,000 Bq. Similarly:

After three half-lives: Activity = 2,000 Bq.

After four half-lives: Activity = 1,000 Bq.

Thus, the rod has halved its activity a minimum of 4 times in 300 days. Therefore, the largest possible estimate for the half-life is given by:

300/4 = 75 days.

Question 67: B

There is no change in the atomic mass or proton numbers in gamma radiation. In alpha decay, a particle with two protons and two neutrons is emitted. This results in a decrease in proton number by 2 and neutron number by 2.

Thus, after 3 rounds of alpha decay, the proton number will be 89 − (3 × 2) = 83. Initially, there are 200 − 89 = 111 neutrons. After the alpha decay rounds, the number of neutrons is 111 − (3 × 2) = 105.

Question 68: C

The speed of the wave can be calculated by using the distance and time given:

$$\text{Speed} = \frac{\text{distance}}{\text{time}} = \frac{500}{1.5} = 333 \text{ ms}^{-1}.$$

As the wavelength is linked to wave speed, we can utilise this equation:

$$\text{Wavelength} = \frac{\text{Speed}}{\text{Frequency}} = \frac{333}{440}.$$

If you approximate 333 to 330, it is easier to reduce the fraction:

$$\frac{330}{440} = \frac{3}{4} = 0.75 \text{ m}.$$

Question 69: B

Firstly, it should be noted that all the options are a magnitude of 10 apart. Therefore, you do not have to worry about getting the correct value as long as you get the correct power of 10. You can therefore make your life easier by approximating.

The area of the shell is given by the following as it has a circular face:

$A_{shell} = \pi r^2 = \pi \times (50 \times 10^{-3})^2 = \pi \times (5 \times 10^{-2})^2 = \pi \times 25 \times 10^{-4} = 7.5 \times 10^{-3} \, m^2$

The deceleration of the shell is easy to evaluate using the following equation:

$a = \dfrac{u-v}{t} = \dfrac{200}{500 \times 10^{-6}} = 0.4 \times 10^6 \, ms^{-2}$.

Then, using Newton's Second Law:

Braking force $=$ Mass $\times$ Acceleration $= 1 \times (0.4 \times 10^6) = 4 \times 10^5 \, N$.

Therefore, we can now calculate the pressure:

Pressure $= \dfrac{Force}{Area} = \dfrac{4 \times 10^5}{7.5 \times 10^{-3}} = \dfrac{8}{15} \times 10^8 \, Pa \approx 5 \times 10^7 Pa$.

Question 70: B

The fountain transfers 10% of 1,000 J of energy per second into lifting the 120 litres of water per minute. Thus, it transfers 100 J into 2 litres of water per second. Therefore, we can equate this to the gravitational energy equation:

Total Gravitational Potential Energy, $E_p = mg\Delta h$

$\Rightarrow 100 \, J = 2 \times 10 \times h \Rightarrow h = \dfrac{100}{20} = 5 \, m$.

Question 71: E

In step-down transformers, the number of turns of the primary coil is larger than that of the secondary coil to decrease the voltage. If a transformer is 100% efficient, the electrical power input is equal to the electrical power output (P = IV). Therefore, both statements 1 and 3 are true.

Question 72: C

The percentage of C^{14} in the bone halves every 5,730 years. Since it has decreased from 100% to 6.25%, it has undergone 4 half-lives – this can be seen by continuously halving 100. On the fourth iteration, 6.25 % is achieved, therefore there have been 4 half-lives during this period. Therefore, the bone is 4 × 5,730 years old = 22,920 years.

Question 73: E

This is a straightforward question in principle, as it just requires you to plug the values into the following equation - just ensure you work in SI units to get the correct answer.

Velocity $=$ Wavelength $\times$ Frequency.

$\Rightarrow$ Frequency $= \dfrac{2 \, m/s}{2.5 \, m} = 0.8 \, Hz = 0.8 \times 10^{-6} \, MHz = 8 \times 10^{-7} \, MHz$.

Question 74: E

If an element has a half-life of 25 days, its population will be halved every 25 days.

A total of 350/25 = 14 half-lives have elapsed. Thus, the count rate has halved 14 times. Therefore, to calculate the original rate, the final count rate must be doubled 14 times = 50×2^{14}.

$2^{14} = 2^5 \times 2^5 \times 2^4 = 32 \times 32 \times 16 = 16,384$.

Therefore, the original count rate is given by:

Original count rate = $50 \times 2^{14} = 16,384 \times 50 = 819,200$.

Question 75: D

Recall the following relations for voltage and power:

$$\text{Voltage} = \text{Current} \times \text{Resistance} = \frac{\text{Power}}{\text{Current}},$$

$$\text{Power} = \frac{\text{Work Done}}{\text{Time}} = \frac{\text{Force} \times \text{Distance}}{\text{Time}} = \text{Force} \times \text{Velocity}.$$

The first relation, $V = IR$, can directly be used to derive A. The second relation, $V = \frac{P}{I}$, and the final equation for power, can be used to derive B. C is derived from: $\text{Voltage} = \frac{\text{Power}}{\text{Current}} = \frac{\text{Force} \times \text{Velocity}}{\text{Current}}$.

As charge is the product of the current and time, E and F are derived from:

$$\text{Voltage} = \frac{\text{Power}}{\text{Current}} = \frac{\text{Force x Distance}}{\text{Time x Current}} = \frac{J}{As} = \frac{J}{C}.$$

D is incorrect as NmC = JC which, by the expression above is not equal to the volt. By comparing this to the expression, we can see that the correct variant would instead be NmC^{-1}.

Question 76: E

The forces acting on the ball are weight, which is constant and equal to the mass times the gravitational acceleration, and tension T which varies with position. By substituting the expression for acceleration given into Newton's Second Law, we obtain:

$$F = ma, \quad a = \frac{v^2}{r}$$

$$\Rightarrow T + mg = m\frac{v^2}{r}$$

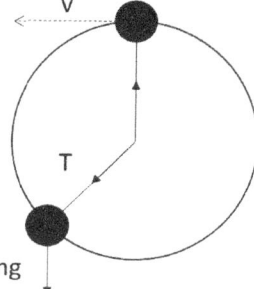

The minimum speed at the top of the arc is when it just manages to reach the top, which is when the tension is completely provided by the weight. Therefore, the force keeping the object moving in the circle is instantaneously its weight. Therefore:

$$T = 0 \Rightarrow mg = m\frac{v^2}{r} \Rightarrow v = \sqrt{gr}.$$

Question 77: E

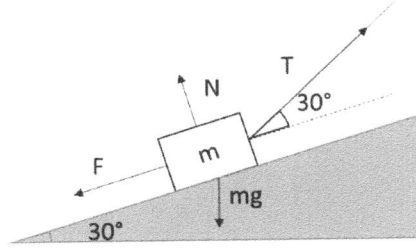

To move at a steady velocity, there must be zero acceleration, so the forces on the object must be balanced. By resolving the forces along the slope, we obtain the following equation:

$$F + mg\sin(30) = T\cos(30) \Rightarrow \frac{mg}{2} + F = \frac{T\sqrt{3}}{2}$$

$$\therefore T = \frac{2}{\sqrt{3}}\left(\frac{mg}{2} + F\right)$$

Work done in pulling the box is given by the relation W = F × d, where the distance can be expressed as a function of velocity and time, $d = v\Delta t$, so:

$$W = vF\Delta t$$

Since power can expressed as $P = \frac{W}{\Delta t}$, we have that $P = \frac{vF\Delta t}{\Delta t}$ and we can rearrange as follows:

$$P = Fv \Rightarrow P = vT\cos(30) \Rightarrow P = \left(\frac{mg}{2} + F\right)v.$$

Question 78: C

This is a conservation of energy problem. In the absence of friction, there is no dissipation of energy – therefore, the sum of the potential and kinetic energy must be constant: $\frac{1}{2}mv^2 + mgh = E$.

At its highest point, the velocity and kinetic energy are both zero so $E = mgh_1$.

At the bottom of the swing, the potential energy is converted to kinetic energy. Therefore:

$$\frac{1}{2}mv^2 = E = mgh_1 \therefore v = \sqrt{2gh_1}$$

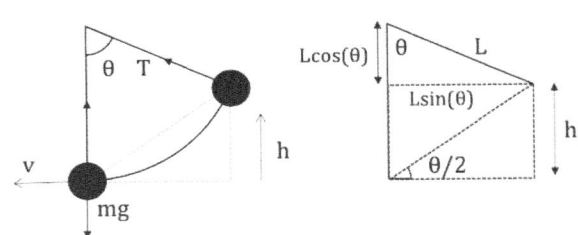

From the diagrams on the left, this initial height can easily be found:

$$h_1 = l(1 - \cos(\theta))$$

$$\Rightarrow v = \sqrt{2gh_1} = \sqrt{2gl(1 - \cos(\theta))}$$

Question 79: A

As the light intensity increases, the resistance of the LDR will decrease. As the potential difference is proportional to the resistance, according to Ohm's Law, this will increase the share of the potential across the normal resistor. Therefore, V_{out} will ultimately increase.

Question 80: D

This is essentially a modified Newton's pendulum. An elastic collision is one in which both kinetic energy and momentum are conserved. The options provided indicate that the balls which move will have the velocity. Therefore, we can use this information to answer the question.

Using the conservation of momentum, we can see that if three balls move after the collision, then:

$$p_i = (3m)v, \ p_f = mu + mu + mu = 3mu$$

$$\therefore 3mv = 3mu \ \Rightarrow \ v = u.$$

Therefore, D is a possible solution. However, if there had been 1 ball moving after the collision, then A is also clearly possible from the conservation of momentum. Therefore, we have to use the kinetic energy conservation to determine the answer. This has been done, using the u = v result above for three bodies:

$$\frac{1}{2}(3m)v^2 = \frac{1}{2}mv^2 + \frac{1}{2}mv^2 + \frac{1}{2}mv^2 = \frac{3}{2}mv^2$$

Therefore, the kinetic conservation is indeed conserved in this case. This can be shown to not be true for A using an identical calculation. Therefore, three balls will swing at a velocity equal to the velocity of the first to conserve both momentum and kinetic energy.

N.B – although calculating from first principles would not be expected, it is possible to get the answer by checking which of the answer choices agree with the conservation of momentum and energy.

Question 81: E

The displacement is the difference between the final position and the initial position. The distance, however, is the total length of the path travelled by the ball. Therefore, the displacement will be zero, as the initial and final position are the same. For the distance, SUVAT or energy conservation may be used. Using energy conservation, the kinetic energy equals the gravitational potential energy.

$$\tfrac{1}{2}mv^2 = mgh \Rightarrow h = \tfrac{1}{2g}(5)^2 = \tfrac{1}{2} \times 2.5 = 1.25 \text{ m}$$

The total distance will be twice this as the object traverses the path again during its descent. Therefore, the final answer is 1.25 × 2 = 2.5 m for the distance.

Question 82: C

The centre of the mass of the two blocks will be at their geometric centres. So, in this case, that is halfway across each rectangle. The moments can be taken about the black point indicated on the diagram, where the weight of the blocks are used:

1 × F = (1 × (20 × 10)) + (2.5 × (20 × 10)) = 700 N.

This corresponds to a mass of 70 kg and so this will be shown on the scale.

Question 83: C

At the top of the bounce, the kinetic energy is zero as velocity is zero. Highest velocity will be downwards before impact, where the potential energy lost is equal to the kinetic energy gained (assuming no air resistance, the conservation of energy can be applied).

Therefore, the kinetic energy before hitting the ground is given by:

$$\frac{1}{2}mv^2 = mgh \Rightarrow E_k = m \times 10 \times 3 = 30m$$

The highest velocity is given by:

$$v^2 = 2gh = 60 \Rightarrow v = 2\sqrt{15}.$$

Question 84: C

A body can only be in equilibrium if all the forces are parallel or they all pass through one point, so 1 and 3 fulfil this. Both 2 and 4 cannot be in equilibrium as the balances are not balanced regardless of the magnitude of the forces involved.

Question 85: B

The initial kinetic energy must equal the work done, by the braking force, to stop the car. Therefore, we can equate the following, where the first step makes use of the fact that the force is half the weight of the car:

$$\frac{1}{2}mv^2 = Fd \Rightarrow \frac{1}{2}mv^2 = \frac{mg}{2}d$$
$$\therefore v^2 = gd \Rightarrow d = \frac{v^2}{g}.$$

Question 86: D

We can use the proportion of amplitude left to work out how many half-lives, the time taken for the amplitude to half, have passed:

$$\frac{25}{200} = \frac{1}{8} = \frac{1}{2^3}.$$

This suggests that 3 half-lives have passed in this duration. Therefore, 12 seconds is three half-lives and $t_{1/2}$ = 12/3 = 4s.

Question 87: B

The question asks for the power dissipated – this occurs due to the frictional force. The equation P = Fv will be utilised, where F is the frictional force. The normal force is *mg*, so the frictional force is given by *μmg*. The power, therefore, is *μmgv*. The acceleration is a red herring here.

Question 88: E

The two waves would interfere destructively as they are half a wavelength phase difference. A wave would reflect back onto itself in this way if reflected from a plane, perpendicular surface. These two waves travelling in opposite directions (incident and reflected) would produce a standing wave, with this exact point in time corresponding to zero amplitude. There are 5 nodes with two fixed ends making it the 4th harmonic of a standing wave. Thus, all the statements are true.

Question 89: C

Beta decay changes a neutron to a proton – therefore, a doesn't change but b increases by one. Then, in alpha decay, the product emits an alpha particle which is two protons and two neutrons - therefore, a decreases by 4 and b decreases by two. This results in a decrease of 4 for a, and 1 for b.

Question 90: A

Use a free body diagram of half the sphere, essentially cutting it down the middle, as shown on the right. The p indicates the internal fluid pressure.

The pressure force acting against the plane is uniform and is given by:

$F = P \times A = P \times \pi r^2$

The stress is uniform around the circumference – where the area of the strip is $2\pi r t$ – it is given by:

Stress force = $\sigma \, 2\pi r t$

By setting the two terms equal for equilibrium, you immediately obtain the answer A.

Question 91: C

Springs behave like capacitors in series or parallel (the opposite of resistors), so we find the effective spring constant of the springs in series to be: $\frac{1}{k} + \frac{1}{k} = \frac{k^2}{2k} = \frac{k}{2}$. Then, we can add the constant of the parallel spring to obtain $\frac{3k}{2}$.

Question 92: A

A free-body diagram on the trailer gives a force exerted by the spring, causing the trailer to accelerate at 2 ms^{-2}. The force is given by:

$F = ma = 10{,}000 \times 2 = 20{,}000$ N.

As this is the force exerted by the spring, it is also equal to:

$F = kx \Rightarrow x = F/k = 20{,}000/100{,}000 = 0.2$ m.

Therefore, the energy stored in the spring is given by:

$E = \frac{1}{2} kx^2 = \frac{1}{2} \times 100{,}000 \times (0.2)^2 = 2000$ J.

Question 93: C

The current in the circuit can be determined using the power and the EMF. This gives a value of:

$$P = IV \Rightarrow I = \frac{P}{V} = \frac{10}{5} = 2 \text{ A}.$$

This can be used to determine the number of electrons passing through the LED in 10 seconds:

$$Q = It = 2 \times 10 = 20 \text{ C} \Rightarrow N = 20/(1.6 \times 10^{-19}) \approx \frac{200}{16} \times 10^{19} \approx 1.25 \times 10^{20}$$

The energy of each electron is easily determined using:

$$W = Vq = 5 \times (1.6 \times 10^{-19}) = 8 \times 10^{-19} \text{ J}.$$

Question 94: B

Moments taken with the pivot at the wall must balance. Therefore, if we label the distance from the wall to the flower pot as L, we can write:

$$\frac{2}{3} LT \sin\theta = Lmg \qquad \Rightarrow T = \frac{3mg}{2 \sin\theta}$$

Question 95: A

The superposed signal will have a frequency of 150 kHz, the lowest common factor of 30 kHz and 50 kHz. Therefore, the time period is given by:

$$T = 1/f = 6.67 \text{ μs}.$$

Question 96: B

The Young's Modulus, E, is from the gradient before the elastic limit, which is readily evaluated from the graph:

$$E = \frac{\Delta \, Stress}{\Delta \, Strain} = \frac{500 \text{ MPa}}{0.05} = 10 \text{ GPa}.$$

The strain energy is evaluated from the area beneath the graph up to x, the elastic limit:

$$\text{Strain Energy} = \frac{1}{2} \times 0.05 \times 500 \text{ MPa} = 12.5 \text{ MJ}.$$

Question 97: D

The impulse is the change in momentum. Therefore, we can use the impulse and the mass to determine the change in velocity of the ball:

$$I = \Delta mv \Rightarrow \Delta v = \frac{I}{m} = \frac{0.27}{0.1} = 2.7 \text{ m/s}.$$

Question 98: C

This problem relies on knowledge of projectiles and can be derived by differentiating an expression for the distance in terms of θ. The velocity component in the horizontal direction is vcosθ. Therefore:

Speed = Distance / Time ⇒ Distance = Speed × Time = vcosθ × t

The time, however, is obtained using SUVAT. This can be substituted to give:

Distance = Speed × Time = vcosθ × $\frac{2v\sin\theta}{g}$ = $\frac{v^2\sin2\theta}{g}$.

This is maximised at 45 degrees.

Question 99: C

Only 90% of the motor's power is used to provide the driving force on the car. Therefore, we can use this and the equation which relates power to speed to give:

Power = 0.9P ⇒ P = Fv ⇒ F=0.9P/v

Given the force, we can easily calculate the work done by the resistive force using d = 1000m:

W = Fd ⇒ W = 900P/v.

Question 100: C

There will be an inertial force on the block. A free body diagram reveals a vertical force mg and horizontal force ma acting on the block. The components of these must balance in the direction parallel to the wedge.

By equating the component of weight parallel to the wedge and the component of the driving force parallel to the wedge, we obtain:

ma cos θ = mg sin θ ⇒ a = g tan θ.

Question 101: B

Each three-block combination is mutually exclusive to any other combination, so the probabilities are added. Each block pick is independent of all other picks, so the probabilities can be multiplied. For this scenario, there are three possible combinations:

P(2 red blocks and 1 yellow block) = P(Red, red, yellow) + P(Red, yellow, red) + P(Yellow, red, red)

∴ P (2 red blocks and 1 yellow block) = $(\frac{12}{20} \times \frac{11}{19} \times \frac{8}{18})$ + $(\frac{12}{20} \times \frac{8}{19} \times \frac{11}{18})$ + $(\frac{8}{20} \times \frac{12}{19} \times \frac{11}{18})$

⇒ P (2 red blocks and 1 yellow block) = $\frac{3 \times 12 \times 11 \times 8}{20 \times 19 \times 18}$ = $\frac{44}{95}$.

Question 102: C
This is simple algebraic manipulation. Multiply through by 15, then rearrange for *x*:

$3(3x + 5) + 5(2x - 2) = 18 \times 15$

$\Rightarrow 9x + 15 + 10x - 10 = 270$

$\Rightarrow 9x + 10x = 270 - 15 + 10 \Rightarrow 19x = 265$

$\therefore x \approx 13.95$.

Question 103: C
This is a rare case where you need to factorise a complex polynomial. The first term is clearly the product of 3x and x, therefore, we can insert these into the brackets immediately. Now, we can consider possible pairings with sum to give 11 and produce a product of -20.

$(3x + a)(x + b) = 0$, possible pairs: $2 \times 10, 10 \times 2, 4 \times 5, 5 \ 4$

$\therefore (3x - 4)(x + 5) = 0$

$\Rightarrow 3x - 4 = 0$, so $x = \frac{4}{3}$.

$\Rightarrow x + 5 = 0$, so $x = -5$.

Therefore, these are the two possible solutions for x.

Question 104: C
This question just requires some basic algebraic manipulation – make sure you don't make silly mistakes in these questions! The steps are as follows:

$\frac{5(x-4)}{(x+2)(x-4)} + \frac{3(x+2)}{(x+2)(x-4)}$

$= \frac{5x-20+3x+6}{(x+2)(x-4)} = \frac{8x-14}{(x+2)(x-4)}$.

Question 105: E
As *p* is directly proportional to the cube root of *q*, we can write the following relation, where k is the constant of proportionality:

$p \propto \sqrt[3]{q} \Rightarrow p = k\sqrt[3]{q}$

Therefore, we can use the fact that when p = 12, q = 27 to determine the value of k.

$12 = k(\sqrt[3]{27}) = 3k \Rightarrow k = 4$.

Now, at p = 24, we can simply substitute in the values for k and p to determine q:

$p = 4\sqrt[3]{q} \Rightarrow 24 = 4\sqrt[3]{q}$,

$\therefore 6 = \sqrt[3]{q} \Rightarrow q = 6^3 = 216$.

Question 106: A

As we are looking for the prime factors of 72^2, we just need to find the prime factors of 72 and then square them. This is done as follows:

$8 \times 9 = 72$

$\Rightarrow 8 = (4 \times 2) = 2 \times 2 \times 2 = 2^3$

$\Rightarrow 9 = 3 \times 3 = 3^2$

$\therefore (2^3 \times 3^2)^2 = 2^6 \times 3^4$

Question 107: C

Note that $1.151 \times 2 = 2.302$. Therefore, we can divide the numerator and denominator of the expression by 1.151, which produces:

$$\frac{2 \times 10^5 + 2 \times 10^2}{10^{10}} = 2 \times 10^{-5} + 2 \times 10^{-8}$$

$\Rightarrow 0.00002 + 0.00000002 = 0.00002002.$

Question 108: E

We can simply expand the expression given, and compare the coefficients to determine the values of **a** and **b**. This is done as follows:

$(y + 2)^2 - 5 = y^2 + 4y + 4 - 5 = y^2 + 4y + 4 - 5 = y^2 + 4y - 1.$

$\therefore y^2 + ay + b = y^2 + 4y - 1.$

Therefore, by directly equating the coefficients, we can see that: **a** = 4 and **b** = -1.

Question 109: E

To simplify the expression, take $5(m + 4n)$ as a common factor to give:

$$\frac{4(m+4n)}{5(m+4n)} + \frac{5(m-2n)}{5(m+4n)}.$$

This can be further simplified to give:

$$\frac{4m+16n+5m-10n}{5(m+4n)} = \frac{9m+6n}{5(m+4n)} = \frac{3(3m+2n)}{5(m+4n)}.$$

Question 110: C

As A is inversely proportional to the square root of B, we can write the following relation, where k is the constant of proportionality:

$$A \propto \frac{1}{\sqrt{B}} \Rightarrow A = \frac{k}{\sqrt{B}}.$$

Therefore, we can use the fact that when A = 4, B = 25 to determine the value of k. Substitute the values in to give:

$$4 = \frac{k}{\sqrt{25}} \Rightarrow k = 4 \times 5 = 20.$$

$$\therefore A = \frac{20}{\sqrt{B}}.$$

So, when B = 16, A is given by:

$$A = \frac{20}{\sqrt{16}} = \frac{20}{4} = 5.$$

Question 111: E

This question tests your knowledge of circle theorems. Angles SVU and STU are opposites and add up to 180° - therefore, STU = 91°.

The angle of the centre of a circle is twice the angle at the circumference - therefore, SOU = 2 × 91° = 182°.

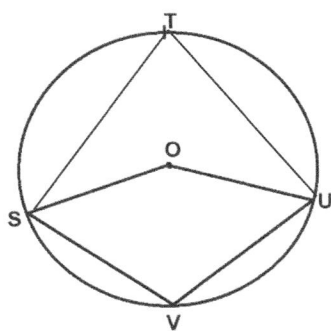

Question 112: E

Cylinder B is an enlargement of A, so the increases in radius (r) and height (h) will be proportional: $\frac{r_A}{r_B} = \frac{h_A}{h_B}$. Let us call the constant of proportionality n, where $n = \frac{r_A}{r_B} = \frac{h_A}{h_B}$. Therefore, we can write the following, where the surface area of the open cylinder is given by A = 2πrh:

$$\Rightarrow \frac{\text{Area A}}{\text{Area B}} = \frac{2\pi r_A h_A}{2\pi r_B h_B} = n \times n = n^2 \Rightarrow \frac{\text{Area A}}{\text{Area B}} = \frac{32\pi}{8\pi} = 4 \therefore n = 2.$$

The constant of proportionality, n = 2, also applies to the volume, where it must be accounted for three times due to the three lengths. The cylinder's volumes are therefore related by $n^3 = 8$.

If the smaller cylinder has volume 2π cm³, then the larger will have volume 2π × n^3 = 2π × 8 = 16π cm³.

Question 113: E

This is another simple algebraic manipulation question. The steps are as follows, where we have cross-multiplied to obtain a common denominator:

$$\Rightarrow \frac{8}{x(3-x)} - \frac{6(3-x)}{x(3-x)} \Rightarrow \frac{8 - 18 + 6x}{x(3-x)} = \frac{6x - 10}{x(3-x)}$$

Question 114: B

For the black ball to be drawn in the last round, white balls must be drawn every round. Therefore, the probability of obtaining 9 white balls in a row is given by:

$$P = \frac{9}{10} \times \frac{8}{9} \times \frac{7}{8} \times \frac{6}{7} \times \frac{5}{6} \times \frac{4}{5} \times \frac{3}{4} \times \frac{2}{3} \times \frac{1}{2}.$$

$$\Rightarrow P = \frac{9 \times 8 \times 7 \times 6 \times 5 \times 4 \times 3 \times 2 \times 1}{10 \times 9 \times 8 \times 7 \times 6 \times 5 \times 4 \times 3 \times 2 \times 1} = \frac{1}{10}.$$

Question 115: C

There are 4 Kings within a standard deck. Therefore, the probability of obtaining a King is $\frac{4}{52} = \frac{1}{13}$, and the probability of obtaining a King again, after the first card, is given by $\frac{3}{51}$. These are independent events – therefore, the probability of drawing two Kings is their product:

$$P = \frac{1}{13} \times \frac{3}{51} = \frac{3}{663} = \frac{1}{221}.$$

Question 116: B

The probabilities of all outcomes must sum to one, so if the probability of rolling a 1 is x, then:

$$\Rightarrow x + x + x + x + 2x = 1 \quad \therefore x = \frac{1}{7}.$$

The probability of obtaining two sixes, therefore, is given by:

$$P_{12} = \frac{2}{7} \times \frac{2}{7} = \frac{4}{49}.$$

Question 117: B

There are plenty of ways of obtaining the answer, however the easiest is as follows: 0 is divisible by both 2 and 3. Half of the numbers from 1 to 36 are even (i.e. 18 of them). 3, 9, 15, 21, 27, 33 are the only numbers divisible by 3 that we've missed in our even numbers. Therefore, there are 25 outcomes divisible by 2 or 3, out of 37.

Question 118: C

To approach this question, list the six ways of achieving this outcome: HHTT, HTHT, HTTH, TTHH, THTH and THHT. The probability of each of these six outcomes is given by:

$$\frac{1}{2} \times \frac{1}{2} \times \frac{1}{2} \times \frac{1}{2} = \left(\frac{1}{2}\right)^4 = \frac{1}{2^4}$$

Therefore, the probability of two heads and two tails is:

$$P = 6 \times \frac{1}{2^4} = \frac{6}{16} = \frac{3}{8}.$$

Question 119: D

Firstly, we need to count the number of ways to get a 5, 6 or 7 (draw the square below if helpful). The ways to get a 5 are: 1, 4; 2, 3; 3, 2; 4, 1. The ways to get a 6 are: 1, 5; 2, 4; 3, 3; 4, 2; 5, 1. The ways to get a 7 are: 1, 6; 2, 5; 3, 4; 4, 3; 5, 2; 6, 1. That is 15 out of 36 possible outcomes, which reduces to 5 out of 12.

	1	2	3	4	5	6
1	2	3	4	5	6	7
2	3	4	5	6	7	8
3	4	5	6	7	8	9
4	5	6	7	8	9	10
5	6	7	8	9	10	11
6	7	8	9	10	11	12

Question 120: C

In total, there are $x + y + z$ balls in the bag, and the probability of picking a red ball is $\frac{x}{(x+y+z)}$ and the probability of picking a green ball is $\frac{z}{(x+y+z)}$. These are independent events, so the probability of picking red then green is the product of their individual probabilities: $\frac{xz}{(x+y+z)^2}$.

The probability of picking green then red is the same. These outcomes are mutually exclusive, so are added. This gives an answer of $\frac{2xz}{(x+y+z)^2}$.

Question 121: B

There are two ways of obtaining a red ball and a blue ball: pulling out a red ball then a blue ball, or pulling out a blue ball and then a red ball. Let us work out the probability of the first:

$$P = \frac{x}{(x+y+z)} \times \frac{y}{(x+y+z-1)}$$

The probability of the second option will be the same. These are mutually exclusive options, so the probabilities may be summed. This gives the answer $\frac{2xy}{(x+y+z)(x+y+z-1)}$.

Question 122: A

Let x correspond to when Player 1 wins the point, and y correspond to when Player 2 wins the point.

Player 1 wins the game in five rounds in the following scenarios: *yxxxx, xyxxx, xxyxx, xxxyx*. (Note the case of *xxxxy* would lead to player 1 winning in 4 rounds, which the question forbids.)

Each of these have a probability of $p^4 \times (1-p)$. Thus, the solution is this multiplied by four as there are four ways for Player 1 to win in 5 rounds; therefore, the final answer is $4p^4(1-p)$.

Question 123: F

This question requires some simple algebraic manipulation, as follows:

$$4x + 7 + 18x + 20 = 14$$

$$\Rightarrow 22x + 27 = 14 \Rightarrow 22x = -13$$

$$\therefore x = -\frac{13}{22}.$$

Question 124: D

Rearrange the expression for volume to obtain an expression for r:

$$r^3 = \frac{3V}{4\pi} \Rightarrow r = \left(\frac{3V}{4\pi}\right)^{1/3}$$

We can substitute this expression into the equation for surface area to find a relationship between S and V.

$$S = 4\pi\left[\left(\frac{3V}{4\pi}\right)^{\frac{1}{3}}\right]^2 = 4\pi\left(\frac{3V}{4\pi}\right)^{\frac{2}{3}}$$

$$= \frac{4\pi(3V)^{\frac{2}{3}}}{(4\pi)^{\frac{2}{3}}} = (3V)^{\frac{2}{3}} \times \frac{(4\pi)^1}{(4\pi)^{\frac{2}{3}}}$$

$$= (3V)^{\frac{2}{3}}(4\pi)^{1-\frac{2}{3}} = (4\pi)^{\frac{1}{3}}(3V)^{\frac{2}{3}}.$$

Question 125: A

Let one side of the cube be of length x. Therefore:

$$S = 6x^2 \Rightarrow x = \left(\frac{S}{6}\right)^{\frac{1}{2}}$$

We ca substitute this expression into the equation for V to find the relation between S and V:

$$\Rightarrow V = x^3 \Rightarrow V = \left[\left(\frac{S}{6}\right)^{\frac{1}{2}}\right]^3$$

$$\therefore V = \left(\frac{S}{6}\right)^{\frac{3}{2}}$$

Question 126: B

By multiplying the second equation by 2, we obtain $4x + 16y = 24$. Subtracting the first equation from this, we get:

$$13y = 17 \Rightarrow y = \frac{17}{13}.$$

Then, by substituting the expression for y into the first equation, we obtain $x = \frac{10}{13}$. You could also try substituting possible solutions one by one, although given that the equations are both linear and contain easy numbers, it is quicker to solve them algebraically.

Question 127: A

Firstly, we multiply by the denominator and partially expand the brackets on the right side:

$$(7x + 10) = (3y^2 + 2)(9x + 5) \Rightarrow (7x + 10) = 9x(3y^2 + 2) + 5(3y^2 + 2)$$

Next, gather the x terms:

$$7x - 9x(3y^2 + 2) = 5(3y^2 + 2) - 10$$

Take x outside the brackets and rearrange to arrive at the final answer:

$$x[7 - 9(3y^2 + 2)] = 5(3y^2 + 2) - 10$$

$$\therefore x = \frac{5(3y^2 + 2) - 10}{7 - 9(3y^2 + 2)} = \frac{(15y^2)}{(7 - 9(3y^2 + 2))}.$$

Question 128: F

This tests your ability to manipulate indices. The steps are as follows:

$$3x\left(\frac{3x^7}{x^{\frac{1}{3}}}\right)^3 = 3x\left(\frac{3^3 x^{21}}{x^{\frac{3}{3}}}\right) = 3x\,\frac{27x^{21}}{x} = 81x^{21}$$

Question 129: D

The expression given can be rearranged to give:

$$2x \times [2^{\frac{7}{14}} x^{\frac{7}{14}}] = 2x \times [2^{\frac{1}{2}} x^{\frac{1}{2}}]$$

$$= 2x\,(\sqrt{2}\,\sqrt{x}) = 2\left[\sqrt{x}\sqrt{x}\right]\left[\sqrt{2}\sqrt{x}\right] = 2\sqrt{2x^3}.$$

Question 130: A

The equation for the area of a circle is given by: $A = \pi r^2$. Therefore, we can equate the following:

$$10\pi = \pi r^2$$

$$\therefore r = \sqrt{10}$$

The circumference, therefore, is given by:

$$C = 2\pi r = 2\pi\sqrt{10}.$$

Question 131: D

This can be evaluated simply using the rule given:

$3.4 = 12 + (3 + 4) = 19$

$19.5 = 95 + (19 + 5) = 119$

$\therefore (3.4).5 = 119$

Question 132: D

This can be evaluated simply using the rule given:

$2.3 = \dfrac{2^3}{2} = 4$

$4.2 = \dfrac{4^2}{4} = 4$

$\therefore (2.3).2 = 4.$

Question 133: F

This is a tricky question that requires you to know how to 'complete the square'. Alternatively, you could use the quadratic formula. The steps are as follows:

$(x + 1.5)(x + 1.5) = x^2 + 3x + 2.25$

$\Rightarrow (x + 1.5)^2 - 7.25 = x^2 + 3x - 5 = 0$

$\therefore (x + 1.5)^2 = 7.25 = \dfrac{29}{4}$

We can now rearrange this equation for x:

$x + 1.5 = \sqrt{\dfrac{29}{4}} \Rightarrow x = -\dfrac{3}{2} \pm \sqrt{\dfrac{29}{4}} = -\dfrac{3}{2} \pm \dfrac{\sqrt{29}}{2}.$

Question 134: B

Whilst you definitely need to solve this graphically, it is necessary to complete the square for the first equation to allow you to draw it more easily:

$\Rightarrow (x + 2)^2 = x^2 + 4x + 4$

$\therefore y = (x + 2)^2 + 10 = x^2 + 4x + 14$

This is now an easy curve to draw (it is the quadratic $y = x^2$ shifted 2 units left and 10 units up). The turning point of this quadratic is to the left and well above anything in x^3, so the only solution is the first intersection of the two curves in the upper right quadrant around (3.4, 39). Therefore, there is only one intersection.

Question 135: C

By far the easiest way to solve this is to sketch the graphs (don't waste time solving them algebraically). From the graphs, it is evident that $y = 2$ and $y = 1 - x^2$ do not intersect, since the latter has its turning point at $(0, 1)$ and zero points at $x = -1$ and 1. The first two graphs, $y = x$ and $y = x^2$, intersect at the origin and $(1, 1)$, and $y = 2$ runs through both. Therefore, only 3 and 4 do not intersect.

Question 136: B

Notice that you're not required to get the actual values – just the number's magnitude. Thus, 897653 can be approximated to 900,000 and 0.009764 to 0.01. Therefore, the estimate is simply:

$$900,000 \times 0.01 = 9,000.$$

This has an order of magnitude of 10^4 which corresponds to B.

Question 137: C

To begin this, multiply the expression through by 70:

$$7(7x + 3) + 10(3x + 1) = 14 \times 70$$

We can then expand the brackets and simplify:

$$49x + 21 + 30x + 10 = 980$$

$$\Rightarrow 79x + 31 = 980$$

$$\Rightarrow x = \frac{949}{79}.$$

Question 138: A

Firstly, split the equilateral triangle into 2 right-angled triangles and apply Pythagoras' theorem:

$$\Rightarrow x^2 = \left(\frac{x}{2}\right)^2 + h^2 \Rightarrow h^2 = \frac{3}{4}x^2$$

$$\therefore h = \sqrt{\frac{3x^2}{4}} = \frac{\sqrt{3x^2}}{2}$$

The area of a triangle is given by:

$$A = \tfrac{1}{2} \times \text{Base} \times \text{Height} = \frac{1}{2} \times \frac{\sqrt{3x^2}}{2}$$

$$\Rightarrow A = x\frac{\sqrt{3x^2}}{4} = x\frac{\sqrt{3}\sqrt{x^2}}{4} = \frac{x^2\sqrt{3}}{4}.$$

Question 139: A

This is a question testing your ability to spot 'the difference between two squares.' We can factorise the expression to give the following, where the common term has been cancelled out:

$$3 - \frac{7x(5x - 1)(5x+1)}{(7x)^2(5x+1)} \Rightarrow 3 - \frac{(5x - 1)}{7x}.$$

Question 140: C

The easiest way to do this is to 'complete the square':

$(x - 5)^2 = x^2 - 10x + 25$

$\Rightarrow (x - 5)^2 - 125 = x^2 - 10x - 100 = 0$

$\Rightarrow (x - 5)^2 = 125$

$\Rightarrow x - 5 = \pm\sqrt{125} = \pm\sqrt{25}\,\sqrt{5} = \pm 5\sqrt{5}$

$\therefore x = 5 \pm 5\sqrt{5}.$

Question 141: B

Firstly, factorise by completing the square and then simply rearrange for x:

$x^2 - 4x + 7 = (x - 2)^2 + 3$

$\Rightarrow (x - 2)^2 = y^3 + 2 - 3$

$\Rightarrow x - 2 = \pm\sqrt{y^3 - 1}$

$\Rightarrow x = 2 \pm \sqrt{y^3 - 1}.$

Question 142: D

Begin by squaring both sides of the expression, and then rearrange for y:

$(3x + 2)^2 = 7x^2 + 2x + y$

$\Rightarrow y = (3x + 2)^2 - 7x^2 - 2x = (9x^2 + 12x + 4) - 7x^2 - 2x$

$\therefore y = 2x^2 + 10x + 4.$

Question 143: C

This is a fourth order polynomial, which you aren't expected to be able to factorise at GCSE. This is where looking at the options makes your life a lot easier. In all of them, opening the bracket on the right side involves making $(y \pm 1)^4$ on the left side, i.e. the answers are hinting that $(y \pm 1)^4$ is the solution to the fourth order polynomial.

Since there are negative terms in the equations (e.g. $- 4y^3$), the solution has to be:

$(y - 1)^4 = y^4 - 4y^3 + 6y^2 - 4y + 1$

$\Rightarrow (y - 1)^4 + 1 = x^5 + 7$

$\Rightarrow y - 1 = (x^5 + 6)^{\frac{1}{4}}$

$\therefore y = 1 + (x^5 + 6)^{1/4}.$

Question 144: A

Let the width of the television be $4x$ and the height of the television be $3x$. By Pythagoras's theorem:

$$(4x)^2 + (3x)^2 = 50^2 \Rightarrow 25x^2 = 2500$$

$$\therefore x = 10.$$

Therefore, the screen is 30 inches by 40 inches. So, the area is 1,200 inches².

Question 145: C

Firstly, square both sides and then multiply out the brackets:

$$1 + \frac{3}{x^2} = (y^5 + 1)^2$$

$$\Rightarrow \frac{3}{x^2} = (y^{10} + 2y^5 + 1) - 1$$

$$\Rightarrow x^2 = \frac{3}{y^{10}+2y^5} \Rightarrow x = \sqrt{\frac{3}{y^{10} + 2y^5}}.$$

Question 146: C

The easiest method is to double the first equation and triple the second to obtain:

$$\Rightarrow 6x - 10y = 20 \text{ and } 6x + 6y = 39.$$

By subtracting the first equation from the second, we obtain:

$$16y = 19 \Rightarrow y = \frac{19}{16}.$$

By substituting this value of y into the first equation, the value of x can be obtained: $x = \frac{85}{16}$.

Question 147: C

This is fairly straightforward; the first inequality is the easier one to work with. By inserting test values, it is clear to see that B, D and E violate it, so we just need to check A and C in the second inequality.

C: $1^3 - 2^2 < 3$, but A: $2^3 - 1^2 > 3$.

Therefore, the answer is given by C.

Question 148: B

Whilst this can be done graphically, it's quicker to do algebraically (because the second equation is not as easy to sketch). Intersections occur where the curves have the same coordinates.

$$\Rightarrow x + 4 = 4x^2 + 5x + 5$$

$$\Rightarrow 4x^2 + 4x + 1 = 0$$

$$\Rightarrow (2x + 1)(2x + 1) = 0$$

Thus, the two graphs only intersect once at $x = -\frac{1}{2}$.

Question 149: D

It is better to approach this algebraically, as the equations are easy to work with and you would need to sketch very accurately to get the answer. Intersections occur where the curves have the same coordinates.

$$x^3 = x \Rightarrow x^3 - x = 0 \Rightarrow x(x^2 - 1) = 0$$

You need to spot the 'difference between two squares' to obtain:

$$x(x + 1)(x - 1) = 0.$$

Thus, there are 3 intersections - at $x = 0, 1$ and -1.

Question 150: E

Note that the line is the hypotenuse of a right-angled triangle with one side unit length and one side of length ½. By Pythagoras's theorem on the triangle shown in the diagram, we can equate:

$$\left(\frac{1}{2}\right)^2 + 1^2 = x^2 \Rightarrow x^2 = \frac{1}{4} + 1 = \frac{5}{4}$$

$$\Rightarrow x = \sqrt{\frac{5}{4}} = \frac{\sqrt{5}}{\sqrt{4}} = \frac{\sqrt{5}}{2}.$$

Question 151: D

We can eliminate z from equation (1) and (2) by multiplying equation (1) by 3 and adding it to equation (2):

$3x + 3y - 3z = -3$	Equation (1) multiplied by 3.
$2x - 2y + 3z = 8$	Equation (2) then add both equations.
$5x + y = 5$	We label this as equation (4).

Now we must eliminate the same variable z from another pair of equations by using equation (1) and (3):

$2x + 2y - 2z = -2$	Equation (1) multiplied by 2.
$2x - y + 2z = 9$	Equation (3) then add both equations.
$4x + y = 7$	We label this as equation (5).

We now use both equations (4) and (5) to obtain the value of x:

$5x + y = 5$	Equation (4).
$-4x - y = -7$	Equation (5) multiplied by -1.
$x = -2$	

Substitute x back in to calculate y:

$$4x + y = 7 \Rightarrow 4(-2) + y = 7 \Rightarrow -8 + y = 7 \Rightarrow y = 15$$

Substitute x and y back in to calculate z:

$$x + y - z = -1 \Rightarrow -2 + 15 - z = -1 \Rightarrow 13 - z = -1$$

$$-z = -14 \Rightarrow z = 14 \therefore x = -2, y = 15, z = 14.$$

Question 152: D

It is evident that 3a is common to all terms and so, can be factored out to give:

$3a(a^2 - 10a + 25)$ $= 3a(a - 5)(a - 5) = 3a(a - 5)^2.$

Question 153: B

Note that 12 is the LCM (Lowest Common Multiple) of 3 and 4. Thus:

-3 (4x + 3y) = -3 (48)   Multiply each side by -3.

4 (3x + 2y) = 4 (34)   Multiply each side by 4.

-12x – 9y = -144

<u>12x + 8y = 136</u>   Add together.

$\Rightarrow$ -y = -8 $\Rightarrow$ y = 8.

Substitute y back in:

4x + 3y = 48

4x + 3(8) = 48

4x + 24 = 48

4x = 24

$\Rightarrow$ x = 6.

Question 154: E

Don't be fooled, this is an easy question - just obey BODMAS and do not skip steps.

$$\frac{-(25-28)^2}{-36+14} = \frac{-(-3)^2}{-22} \Rightarrow \frac{-(9)}{-22} = \frac{9}{22}.$$

Question 155: E

As there are 26 possible letters for each of the 3 letters in the license plate, and there are 10 possible numbers (0-9) for each of the 3 numbers in the same plate, the number of license plates would be:

$(26) \times (26) \times (26) \times (10) \times (10) \times (10) = 17,576,000.$

Question 156: B

Expand the brackets of the expression given and then factorise:

$\Rightarrow 4x^2 - 12x + 9 = 0 \Rightarrow (2x - 3)(2x - 3) = 0.$

Thus, only one solution exists, $x = 1.5$. Note that you could also use the fact that the discriminant, $b^2 - 4ac = 0$ to get the answer.

Question 157: C

The expression can be rewritten as:

$$\left(x^{\frac{1}{2}}\right)^{\frac{1}{2}} (y^{-3})^{\frac{1}{2}} \Rightarrow y^{-\frac{3}{2}} = \frac{x^{\frac{1}{4}}}{y^{\frac{3}{2}}}.$$

Question 158: A

Let x, y, and z represent the rent for the 1-bedroom, 2-bedroom, and 3-bedroom flats, respectively. We can form 3 different equations using the information given: 1 for the rent, 1 for the repairs, and the last one for the statement that the 3-bedroom unit costs twice as much as the 1-bedroom unit.

(1) x + y + z = 1240

(2) 0.1x + 0.2y + 0.3z = 276

(3) z = 2x

Substitute z = 2x in both of the two other equations to eliminate z:

(4) x + y + 2x = 3x + y = 1240

(5) 0.1x + 0.2y + 0.3(2x) = 0.7x + 0.2y = 276

-2(3x + y) = -2(1240)	Multiply each side of (4) by -2.
10(0.7x + 0.2y) = 10(276)	Multiply each side of (5) by 10.
(6) -6x -2y = -2480	Add these 2 equations.

(7) 7x + 2y = 2760

x = 280

z = 2(280) = 560	Because z = 2x.
280 + y + 560 = 1240	Because x + y + z = 1240.

$\Rightarrow$ y = 400.

Thus, the units rent for £280, £400, £560 per week respectively.

Question 159: C

Following the rules of BODMAS, the expression can be reduced as shown:

$$5\left[5(6^2 - 5 \times 3) + 400^{\frac{1}{2}}\right]^{1/3} + 7$$

$$\Rightarrow 5\,[5(36 - 15) + 20]^{\frac{1}{3}} + 7 \Rightarrow 5\,[5(21) + 20]^{\frac{1}{3}} + 7$$

$$\Rightarrow 5\,(105 + 20)^{\frac{1}{3}} + 7 \Rightarrow (125)^{\frac{1}{3}} + 7 \Rightarrow 5(5) + 7$$

$$\Rightarrow 25 + 7 = 32.$$

Question 160: B

Consider a triangle formed by joining the centre to two adjacent vertices. Six similar triangles can be made around the centre – thus, the central angle is 60 degrees. Since the two lines forming the triangle are of equal length, we have 6 identical equilateral triangles in the hexagon.

Now split the triangle in half and apply Pythagoras' theorem: $1^2 = 0.5^2 + h^2$

Thus, $h = \sqrt{\frac{3}{4}} = \frac{\sqrt{3}}{2}$.

Thus, the area of the triangle is: $\frac{1}{2}bh = \frac{1}{2} \times 1 \times \frac{\sqrt{3}}{2} = \frac{\sqrt{3}}{4}$.

Therefore, the area of the hexagon is: $\frac{\sqrt{3}}{4} \times 6 = \frac{3\sqrt{3}}{2}$.

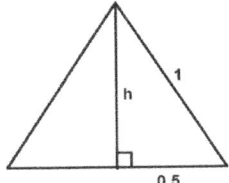 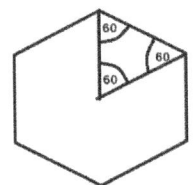

Question 161: B

Let x be the width and $x + 19$ be the length.

Thus, the area of a rectangle is $x(x + 19) = 780$.

Therefore:

$x^2 + 19x - 780 = 0 \Rightarrow (x - 20)(x + 39) = 0$

$\Rightarrow x - 20 = 0$ or $x + 39 = 0$

$\therefore x = 20$ or $x = -39$.

As length cannot be a negative number, we disregard $x = -39$ and use $x = 20$ instead.

Thus, the width is 20 metres and the length is 39 metres.

Question 162: B

The quickest way to solve is by trial and error, substituting the provided options. However, if you're keen to do this algebraically, you can do the following:

Start by setting up the equations:

Perimeter $= 2L + 2W = 34 \Rightarrow L + W = 17 \Rightarrow W = 17 - L$.

Using Pythagoras: $L^2 + W^2 = 13^2$

$\Rightarrow L^2 + (17 - L)^2 = 169$

$\Rightarrow L^2 + 289 - 34L + L^2 = 169$

$\Rightarrow 2L^2 - 34L + 120 = 0$

$\Rightarrow L^2 - 17L + 60 = 0$

$\Rightarrow (L - 5)(L - 12) = 0$

$\therefore L = 5$ and $L = 12$, $W = 12$ and $W = 5$.

Question 163: C

Multiply both sides by 8, expand and then rearrange for x:

$4(3x - 5) + 2(x + 5) = 8(x + 1)$

$\Rightarrow 12x - 20 + 2x + 10 = 8x + 8$

$\Rightarrow 14x - 10 = 8x + 8$

$\Rightarrow 14x = 8x + 18$

$6x = 18 \therefore x = 3$.

Question 164: C

Note that 1.742×3 is 5.226. Therefore, the original equation can be simplified to:

$$\Rightarrow \frac{3 \times 10^6 + 3 \times 10^5}{10^{10}}$$

$= 3 \times 10^{-4} + 3 \times 10^{-5} = 3.3 \times 10^{-4}$.

Question 165: A

The area of a triangle is given by half its base times the height. Therefore, in this case, it is given by:

$$\text{Area} = \frac{(2 + \sqrt{2})(4 - \sqrt{2})}{2} = \frac{8 - 2\sqrt{2} + 4\sqrt{2} - 2}{2}$$

$$\Rightarrow \text{Area} = \frac{6 + 2\sqrt{2}}{2} = 3 + \sqrt{2} \quad .$$

Question 166: C

The aim here is to isolate x, so the first step will be to square both sides, which gives:

$$\frac{4}{x} + 9 = (y - 2)^2 \Rightarrow \frac{4}{x} = (y - 2)^2 - 9$$

Then you cross multiply, followed by factorisation:

$$\frac{x}{4} = \frac{1}{(y-2)^2 - 9} \Rightarrow x = \frac{4}{y^2 - 4y + 4 - 9} \Rightarrow x = \frac{4}{y^2 - 4y - 5}$$

$$\therefore x = \frac{4}{(y+1)(y-5)}.$$

Question 167: D

Set up the equation using the statement provided. This leads to:

$5x - 5 = 0.5(6x + 2)$

$\Rightarrow 10x - 10 = 6x + 2$

$\Rightarrow 4x = 12$

$\therefore x = 3$

Question 168: C

Firstly, you should round the numbers appropriately:

$$\Rightarrow \frac{55 + \left(\frac{9}{4}\right)^2}{\sqrt{900}} = \frac{55 + \frac{81}{16}}{30}$$

As 81 rounds to 80, this can be approximated to give:

$$\frac{55 + 5}{30} = \frac{60}{30} = 2.$$

Question 169: D

There are three outcomes from choosing the type of cheese in the crust. For each of the additional toppings to possibly add, there are 2 outcomes: to include or not to include a certain topping, for each of the 7 toppings.

Thus, the number of different kinds of pizza is:

$3 \times 2 \times 2 \times 2 \times 2 \times 2 \times 2 \times 2 = 3 \times 2^7 = 3 \times 128 = 384.$

Question 170: A

Although it is possible to do this algebraically, by far the easiest way is via trial and error. This is indicated by the fact that rearranging the first equation to make x or y the subject leaves you with a difficult equation to work with (e.g. $x = \sqrt{1 - y^2}$), when you try to substitute it into the second.

An exceptionally good student might notice that the equations are symmetric in x and y, i.e. the solution is when x = y. Thus $2x^2 = 1$ and $2x = \sqrt{2}$ which gives $\frac{\sqrt{2}}{2}$ as the answer.

Question 171: C

If two shapes are congruent, then they are the same size and shape. Thus, congruent objects can be rotations and mirror images of each other. The two triangles in E are indeed congruent (SAS). Congruent objects must, by definition, have the same angles.

Question 172: B

Firstly, rearrange the equation and then factorise the expression:

$x^2 + x - 6 \geq 0$

$\Rightarrow (x + 3)(x - 2) \geq 0$

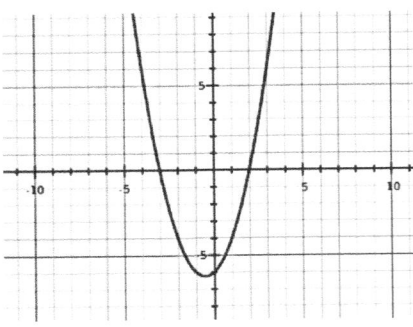

Remember that this is a quadratic inequality, so it requires a quick sketch to ensure you do not make a silly mistake with which way the sign is.

Using the roots of the quadratic, we can see that y = 0 when x = 2 and x = −3. Therefore:

y > 0 when x > 2 or x < −3 $\Rightarrow$ x ≤ −3 and x ≥ 2.

Question 173: B

Using Pythagoras, and the fact that the other two sides are equal in length, we can obtain the following relation between x and the length of the two sides a:

$$a^2 + b^2 = x^2 \Rightarrow a = b \Rightarrow 2a^2 = x^2.$$

$$\text{Area} = \frac{1}{2}\text{base} \times \text{height} = \frac{1}{2}a^2.$$

As $a^2 = \frac{x^2}{2}$, we can rewrite the area in terms of x:

$$\text{Area} = \frac{1}{2} \times \frac{x^2}{2} = \frac{x^2}{4}.$$

Question 174: A

If X and Y are doubled, the value of Q increases by 4. Halving the value of A reduces this to 2. Finally, tripling the value of B reduces this to ⅔, i.e. the value decreases by ⅓. Increases by $\frac{2}{3}$ is incorrect as the expression is scaled by $\frac{2}{3}$, it does not increase by $\frac{2}{3}$ – this would be equivalent to scaling by a factor of by $\frac{5}{3}$.

Question 175: C

The quickest way to do this is to sketch the curves. This requires you to factorise both equations by completing the square:

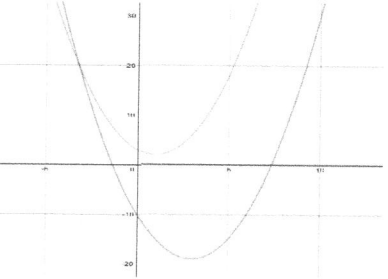

$$x^2 - 2x + 3 = (x-1)^2 + 2 \quad \Rightarrow \quad x^2 - 6x - 10 = (x-3)^2 - 19$$

Thus, the first equation has a turning point at (1, 2) and doesn't cross the x-axis. The second equation has a turning point at (3, -19) and crosses the x-axis twice. The first equation is shown in green on the right, and the second equation is in purple.

Question 176: C

As it is an equilateral triangle, the angles must be equal to 60°. The sector area is given by the following:

$$\theta = 60° \Rightarrow A = \frac{60}{360}\pi r^2 = \frac{1}{6}\pi r^2$$

$$\Rightarrow \frac{x}{\sin 30°} = \frac{2r}{\sin 60°} \quad \therefore x = \frac{2r}{\sqrt{3}}$$

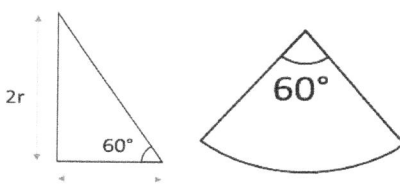

Therefore, the total triangle area is given by:

$$\Rightarrow 2 \times \frac{1}{2} \times \frac{2r}{\sqrt{3}} \times 2r = \frac{4r^2}{\sqrt{3}}.$$

Thus, the proportion covered is given by:

$$\left. \frac{1}{6}\pi r^2 \middle/ \frac{4r^2}{\sqrt{3}} \right. = \frac{\sqrt{3}\pi}{24} \approx 23\%.$$

Question 177: B

To approach this question, construct the triangle shown in the diagram on the right. This allows you to determine the length of the vertical side.

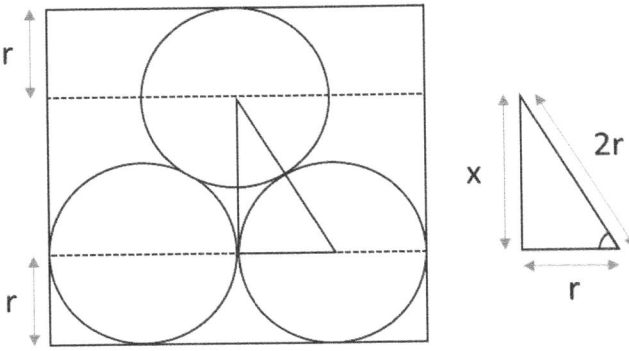

$$\Rightarrow (2r)^2 = r^2 + x^2 \Rightarrow 3r^2 = x^2$$

$$\therefore x = \sqrt{3}r$$

Total height $= 2r + x$

$$\therefore H = (2 + \sqrt{3})r$$

Question 178: A

The volume of a pyramid is given by: $\text{Volume} = \frac{1}{3}\text{Height} \times \text{Base area}$.

Therefore, the base area of both pyramids must be equal if h and V are the same. To find the area of the hexagon, we need the internal angle. This is given by:

External angle $= 360°/6 = 60°$

$\Rightarrow$ Internal angle $= 180° -$ External angle $= 120°$

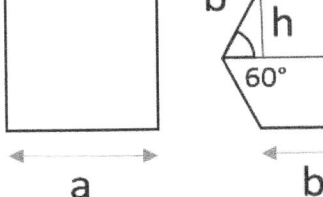

A hexagon can be broken up into two trapezia of height h, as shown in the diagram, where:

$$\frac{b}{\sin 90°} = \frac{h}{\sin 60°} \Rightarrow h = \frac{\sqrt{3}}{2}b$$

$$\therefore \text{Trapezium area} = \frac{(2b+b)}{2}\frac{\sqrt{3}}{2}b = \frac{3\sqrt{3}}{4}b^2$$

$$\therefore \text{Total hexagon area} = \frac{3\sqrt{3}}{2}b^2$$

Now we can equate the areas:

$$a^2 = \frac{3\sqrt{3}}{2}b^2 \Rightarrow \text{Ratio} = \sqrt{\frac{3\sqrt{3}}{2}}.$$

Question 179: C

A cube has 6 sides so the total surface area of a 9 cm cube is given by: 6×9^2.

The 9 cm cube splits into 3 cm cubes. The area of one 3 cm cube is $= 6 \times 3^2$. The total area is 3^3 multiplied by this area. Therefore, the ratio is evaluated as:

$$\Rightarrow \frac{6 \times 3^2 \times 3^3}{6 \times 3^2 \times 3^2} = 3.$$

Question 180: E

The cone is shown in the diagram. To determine θ, trigonometry can be used:

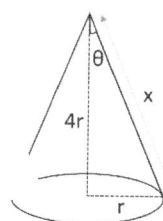

$$\tan \theta = \frac{\text{Opposite}}{\text{Adjacent}} = \frac{r}{4r} = \frac{1}{4}$$

$$\Rightarrow \theta = \tan^{-1}\left(\frac{1}{4}\right).$$

Question 181: C

A hemisphere has an angle of 180 degrees. In this case, this max angle would correspond to the max speed: 200 mph. Therefore, for 70 mph:

$$\frac{\theta}{180} = \frac{70}{200} \quad \Rightarrow \quad \theta = \frac{7 \times 180}{20} = 63°.$$

Question 182: C

Since the two rhombuses are similar, they have identical angles. Therefore, the ratio of their angles is 1.

The length scales with square root of area, therefore:

$$\text{Area}_A = 10 \times \text{Area}_B \Rightarrow \text{Length}_A = \sqrt{10} \times \text{Length}_B$$

$$\Rightarrow \frac{\text{angle A} / \text{angle B}}{\text{length A} / \text{length B}} = \frac{1}{\sqrt{10}/1} = \frac{1}{\sqrt{10}}.$$

Question 183: E

Finding an inverse function is equal to reflecting about the $y = x$ line. Therefore, we can simply interchange the two variables to determine the inverse:

$$y = \ln(2x^2) \Rightarrow e^y = 2x^2 \Rightarrow x = \sqrt{\frac{e^y}{2}}$$

The -x input does not affect the original function, as the value of x is squared. Therefore, $f(-x) = f(x)$. So, the answer remains unchanged.

$$\Rightarrow f(x) = \sqrt{\frac{e^y}{2}}.$$

Question 184: C

Firstly, we can approximate the values provided and then compare then with one another:

$$\Rightarrow \log_8(x) \text{ and } \log_{10}(x) < 0$$

$$\Rightarrow x^2 < 1$$

$$\Rightarrow \sin(x) \leq 1$$

$$\Rightarrow 1 < e^x < 2.72.$$

Therefore, e^x is largest over this range.

Question 185: C

The two relationships given are: $x \propto y^3, y \propto \sqrt{z}$. Therefore, the relationship between x and z is given by:

$$\Rightarrow x \propto \sqrt{z}^3$$

Therefore, if z is doubled, then: $\sqrt{2}^3 = 2\sqrt{2}$.

Question 186: A

The area of the shaded part, that is the difference between the area of the larger and smaller circles, is three times the area of the smaller so: $\pi r^2 - \pi x^2 = 3\pi x^2$. From this, we can see that the area of the larger circle, radius x, must be 4x the smaller one. Therefore:

$$4\pi r^2 = \pi x^2 \Rightarrow 4r^2 = x^2$$

$$\therefore x = 2r$$

The gap, therefore, is given by: $x - r = 2r - r = r$.

Question 187: D

Firstly, rearrange the equation and solve the quadratic:

$$x^2 + 3x - 4 \geq 0 \Rightarrow (x - 1)(x + 4) \geq 0$$

$$\Rightarrow x - 1 \geq 0 \text{ or } x + 4 \geq 0$$

$$\therefore x \geq 1 \text{ or } x \geq -4.$$

Question 188: C

The expressions for the volume of the sphere and its projected area, a circle, are equal:

$$\frac{4}{3}\pi r^3 = \pi r^2 \Rightarrow \frac{4}{3}r = 1$$

$$\therefore r = \frac{3}{4}.$$

Question 189: B

The two graphs have been sketched on the right. Clearly, the two graphs intersect when $x^2 = \frac{1}{x}$, which is at x = 1. Now we can consider the inequality.

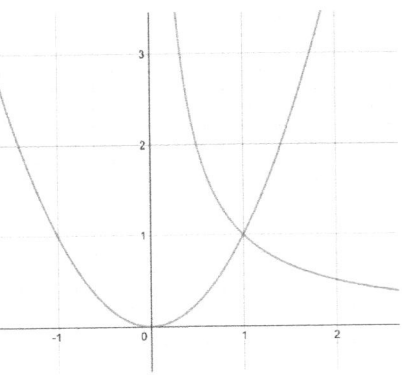

When $x > 1$, $x^2 > 1$, $\frac{1}{x} < 1$.

When $x < 1$, $x^2 < 1$, $\frac{1}{x} > 1$.

In the $x < 0$ range, the $\frac{1}{x}$ graph is negative. Therefore, this region does not satisfy the inequality. Thus, the range is $0 < x < 1$.

Question 190: A

Don't be afraid of how difficult this initially looks. If you follow the pattern, you get (e - e) which equals 0. Anything multiplied by 0 gives zero.

Question 191: C

For two vectors to be perpendicular their scalar product must be equal to 0. Therefore:

$$\begin{pmatrix} -1 \\ 6 \end{pmatrix} \cdot \begin{pmatrix} 2 \\ k \end{pmatrix} = 0 \Rightarrow -2 + 6k = 0$$

$$\therefore k = \frac{1}{3}.$$

Question 192: C

The point q connects the perpendicular line from the plane to the point p.

$$q = -3i + j + \lambda_1(i + 2j)$$

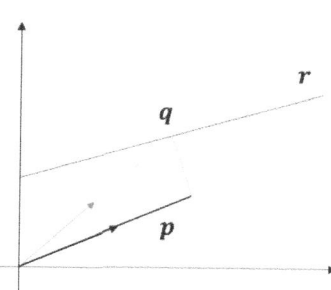

$$\overrightarrow{PQ} = -3i + j + \lambda_1(i + 2j) - 4i - 5j = \begin{pmatrix} -7 + \lambda_1 \\ -4 + 2\lambda_1 \end{pmatrix}$$

PQ is perpendicular to the plane r – therefore, the dot product of $\overrightarrow{PQ}$ and the vector along the line, the direction vector, must be 0. Therefore:

$$\begin{pmatrix} -7 + \lambda_1 \\ -4 + 2\lambda_1 \end{pmatrix} \cdot \begin{pmatrix} 1 \\ 2 \end{pmatrix} = 0 \qquad\qquad \Rightarrow -7 + \lambda_1 - 8 + 4 + \lambda_1 = 0$$

$$\Rightarrow \lambda_1 = 3 \quad \therefore \overrightarrow{PQ} = \begin{pmatrix} -4 \\ 2 \end{pmatrix}$$

The perpendicular distance from the plane to point p is therefore the modulus of the vector joining the two $\overrightarrow{PQ}$:

$$|\overrightarrow{PQ}| = \sqrt{(-4)^2 + 2^2} = \sqrt{20} = 2\sqrt{5}.$$

Question 193: E

This is essentially a system of equations, and you approach it just like a set of simultaneous equations. You obtain these equations by equating each component (x, y and z). There are three equations, and three unknowns, so it is straightforward to solve them.

$$-1 + 3\mu = -7 \quad \therefore \quad \mu = -2$$

$$2 + 4\lambda + 2\mu = 2 \quad \therefore \lambda = 1$$

$$3 + \lambda + \mu = k \quad \therefore \quad k = 2.$$

Question 194: E

Recall the trigonometric identity: $\sin\left(\frac{\pi}{2} - 2\theta\right) = \cos(2\theta)$.

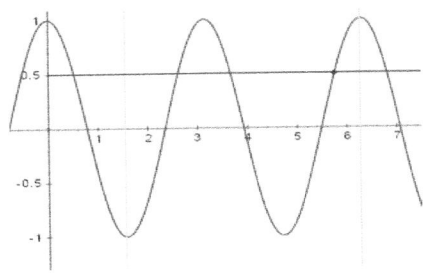

Root solution to $\cos(\theta) = 0.5 \Rightarrow \theta = \frac{\pi}{3}$

Solution to $\cos(2\theta) = 0.5 \Rightarrow \theta = \frac{\pi}{6}$

Largest solution within range is: $2\pi - \frac{\pi}{6} = \frac{(12-1)\pi}{6} = \frac{11\pi}{6}$.

Question 195: A

This requires some knowledge of identities and algebraic manipulation:

$$\cos^4(x) - \sin^4(x) \equiv \{\cos^2(x) - \sin^2(x)\}\{\cos^2(x) + \sin^2(x)\}$$

From difference of two squares, and using Pythagorean identity: $\cos^2(x) + \sin^2(x) = 1$.

$$\Rightarrow \cos^4(x) - \sin^4(x) \equiv \cos^2(x) - \sin^2(x)$$

The double angle formula states that: $\cos(A + B) = \cos(A)\cos(B) - \sin(A)\sin(B)$.

$\therefore$ If $A = B, \cos(2A) = \cos(A)\cos(A) - \sin(A)\sin(A) = \cos^2(A) - \sin^2(A)$.

Therefore, $\cos^4(x) - \sin^4(x) \equiv \cos(2x)$.

Question 196: C

The easiest method here is to insert values and find some factors of the polynomial. This leads to the following factorisation:

$$(x + 1)(x + 2)(2x - 1)(x^2 + 2) = 0$$

Therefore, there are three real roots at $x = -1, x = -2, x = 0.5$ and two imaginary roots at 2i and -2i.

Question 197: C

An arithmetic sequence has constant difference d - so the sum increases by a constant amount, d, each time:

$$u_n = u_1 + (n - 1)d \Rightarrow \sum_{1}^{n} u_n = \frac{n}{2}\{2u_1 + (n - 1)d\}$$

$$\Rightarrow \sum_{1}^{8} u_n = \frac{8}{2}\{4 + (8 - 1)3\} = 100.$$

Question 198: E

The binomial equation needs to be utilised here:

$$\binom{n}{k} 2^{n-k}(-x)^k$$

In this case, n = 5 and k = 2. Therefore, this gives:

$$\Rightarrow \binom{5}{2} 2^{5-2}(-x)^2 = 10 \times 2^3 x^2 = 80x^2.$$

Question 199: A

Having already thrown a 6 is irrelevant. A fair die has equal probability $P = \frac{1}{6}$ for every throw.

For three throws:

$$P(6 \cap 6 \cap 6) = \left(\frac{1}{6}\right)^3 = \frac{1}{216}.$$

Question 200: D

The situation is depicted in the probability tree diagram below:

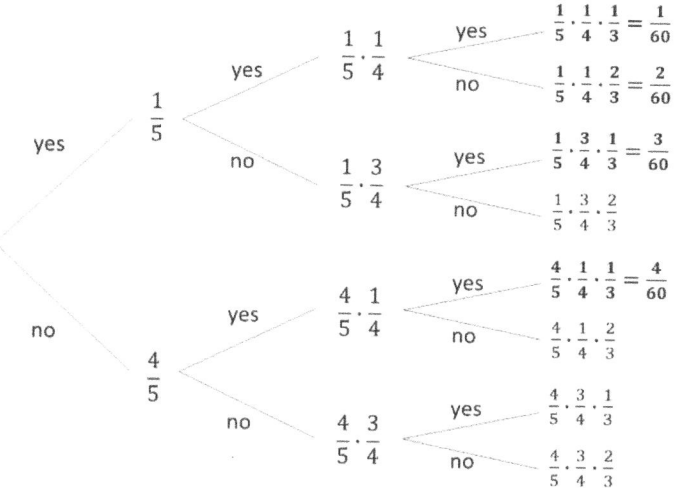

Total probability is sum of all probabilities:

$$P_{\text{Total}} = P(Y \cap Y \cap Y) + P(Y \cap Y \cap N) + P(Y \cap N \cap Y) + P(N \cap Y \cap Y)$$

$$P_{\text{Total}} = \frac{1}{60} + \frac{2}{60} + \frac{3}{60} + \frac{4}{60} = \frac{10}{60} = \frac{1}{6}.$$

Question 201: C

If A represents the probability of having blonde hair, and B the probability of having brown eyes, then we can state the probability of having neither is given by:

$$P[(A \cup B)'] = 1 - P[(A \cup B)] = 1 - \{P(A) + P(B) - P(A \cap B)\} = 1 - \frac{2+6-1}{8} = \frac{3}{8}.$$

Question 202: D

Using the product rule, the following is obtained:

$$\frac{dy}{dx} = x \cdot 4(x+3)^3 + 1 \cdot (x+3)^4 = 4x(x+3)^3 + (x+3)(x+3)^3$$

$$\Rightarrow \frac{dy}{dx} = (5x+3)(x+3)^3.$$

Question 203: C

This is a straightforward definite integral to compute. The steps are as follows:

$$\Rightarrow \int_1^2 \frac{2}{x^2}\,dx = \int_1^2 2x^{-2}\,dx$$

$$\Rightarrow \left[\frac{2x^{-1}}{-1}\right]_1^2 = \left[\frac{-2}{x}\right]_1^2 = \frac{-2}{2} - \frac{-2}{1} = 1.$$

Question 204: D

To express the expression in the desired form, you have to multiply by the complex conjugate, as follows:

$$\frac{5i}{1+2i} \cdot \frac{1-2i}{1-2i}$$

$$\Rightarrow \frac{5i+10}{1+4} = \frac{5i+10}{5} = i + 2.$$

Question 205: B

Firstly, we can use the rules of logarithms to simplify each term:

$$7\log_a(2) - 3\log_a(12) + 5\log_a(3)$$

$$\Rightarrow 7\log_a(2) = \log_a(2^7) = \log_a(128)$$

$$\Rightarrow 3\log_a(12) = \log_a(1728)$$

$$\Rightarrow 5\log_a(3) = \log_a(243)$$

Therefore, the original equation becomes:

$$\log_a(128) - \log_a(1728) + \log_a(243)$$

$$= \log_a\left(\frac{128 \times 243}{1728}\right) = \log_a(18).$$

Question 206: E

Rational functions which are a ratio of two quadratic functions have a horizontal asymptote. This can be determined by dividing each term by the highest order in the polynomial i.e. x^2, as shown:

$$\frac{2x^2 - x + 3}{x^2 + x - 2} = \frac{2 - \frac{1}{x} + \frac{3}{x^2}}{1 + \frac{1}{x} - \frac{2}{x^2}}$$

$$\lim_{x \to \infty}\left(\frac{2 - \frac{1}{x} + \frac{3}{x^2}}{1 + \frac{1}{x} - \frac{2}{x^2}}\right) = \frac{2}{1} \text{ i.e. } y \to 2.$$

So, the equation of the asymptote is $y = 2$.

Question 207: A

The intersection points are found by equating the two expressions:

$1 - 3e^{-x} = e^x - 3$

$\Rightarrow 4 = e^x + 3e^{-x} = \dfrac{(e^x)^2}{e^x} + \dfrac{3}{e^x} = \dfrac{(e^x)^2 + 3}{e^x}$

This is a quadratic equation in e^x:

$(e^x)^2 - 4(e^x) + 3 = 0 \Rightarrow (e^x - 3)(e^x - 1) = 0$

Therefore, $e^x = 3, x = \ln(3)$ or $e^x = 1 \Rightarrow x = 0$.

Question 208: D

To determine the radius, the equation must be rearranged into the format: $(x + a)^2 + (y + b)^2 = r^2$.

$(x - 3)^2 + (y + 4)^2 - 25 = 12$

$\Rightarrow (x - 3)^2 + (y + 4)^2 = 37$

$\therefore r = \sqrt{37}$.

Question 209: C

This question can be attempted graphically or algebraically. For the latter, note that $\sin(-x) = -\sin(x)$. Then the integral becomes:

$\Rightarrow \displaystyle\int_0^a 2\sin(-x)\, dx = -2 \int_0^a \sin(x)\, dx = -2[\cos(x)]_0^a = \cos(a) - 1$

This is equal to zero at the following values:

$\cos(a) - 1 = 0 \therefore a = 2k\pi$.

Therefore, the integral of any whole period of $\sin(x) = 0$, i.e. $a = 2k\pi$, will be 0.

Question 210: E

Firstly, break the expression into the partial fraction terms:

$$\frac{2x + 3}{(x - 2)(x - 3)^2} = \frac{A}{(x - 2)} + \frac{B}{(x - 3)} + \frac{C}{(x - 3)^2}$$

$$\Rightarrow 2x + 3 = A(x - 3)^2 + B(x - 2)(x - 3) + C(x - 2)$$

When $x = 3, (x - 3) = 0 \Rightarrow C = 9$. When $x = 2, (x - 2) = 0 \Rightarrow A = 7$.

$$\Rightarrow 2x + 3 = 7(x - 3)^2 + B(x - 2)(x - 3) + 9(x - 2)$$

Equating coefficients of x^2 on either side gives: $0 = 7 + B$ which gives: $B = -7$.

SECTION 2: WORKED SOLUTIONS

Q	A	Q	A	Q	A	Q	A	Q	A
1.1	C	3.1	B	5.1	C	7.1	E	9.1	C
1.2	D	3.2	E	5.2	B	7.2	D	9.2	B
1.3	C	3.3	B	5.3	A	7.3	B	9.3	A
1.4	C	3.4	D	5.4	C	7.4	D	9.4	C
2.1	A	4.1	B	6.1	E	8.1	C	10.1	C
2.2	E	4.2	D	6.2	A	8.2	D	10.2	E
2.3	D	4.3	C	6.3	B	8.3	A	10.3	D
2.4	A	4.4	A	6.4	A	8.4	D	10.4	A

Question 1.1: C

The distance the ball travels is maximised when the angle to the horizontal at which the golfer hits the ball is 45°. This is shown as follows: the velocity component in the horizontal direction is $v\cos\theta$, where v is the speed of the ball. Therefore:

$$\text{Speed} = \text{Distance} / \text{Time} \Rightarrow \text{Distance} = \text{Speed} \times \text{Time} = v\cos\theta \times t = v\cos\theta \times \frac{2v\sin\theta}{g} = \frac{v^2\sin2\theta}{g}$$

The time, however, is obtained using SUVAT. This expression is maximised at 45 degrees.

Question 1.2: D

The velocity at which the golfer hits the ball is the distance traversed by the club head divided by the time:

$$v = \frac{2\pi L}{T} = 62.5 \text{m/s}.$$

Question 1.3: C

The vertical velocity is $v\sin(45)$, and the acceleration downwards is $g = -10$. The initial velocity is known from the answer of 1.2. Therefore, a SUVAT equation can be used, where the time is twice the time taken for the ball to reach $v = 0$ in the air.

Assuming that the initial velocity of the ball is approximately $v = 60$ m/s, we obtain:

$$v = u + at \Rightarrow t = \frac{v - u}{a} = \frac{0 - 60\sin(45)}{-10} = 3\sqrt{2} \text{ s}$$

As the time t the ball spends in the air is twice the time to reach peak height when $v = 0$, we simply double this value to give a total time of $6\sqrt{2}$ s.

Question 1.4: C

Time spent for vertical motion is the same as that for horizontal. Assuming air resistance is negligible, the distance D the ball travels in that time is then:

$$D = v_h t = v\cos(45°)t = 360 \text{ m}.$$

Question 2.1: A

The mass of the Moon is given by:

$$M_{Moon} = \rho_{Moon} \times V_{Moon} = \rho_{Moon} \times 4r_{Moon}^3 \times \pi/3.$$

As density and radius are different for the Moon that that of Earth, substitute in for $\left(\frac{3}{4}\right)\rho$ and $\left(\frac{1}{4}\right)r$ to make it in terms of the earth's data:

$$M_{Moon} = \frac{3}{4}\rho_{(earth)} \times \frac{4}{3} \times \left(\frac{r_{earth}}{4}\right)^3 \times \pi = \frac{\rho_{earth} \times r_{earth}^3 \times \pi}{64}.$$

Question 2.2: E

The gravitational force is $F = G\frac{Mm}{R^2}$. According to Newton's second law $F = ma$. Therefore, the gravitational acceleration is $= \frac{GM}{R^2}$. Using concepts applied in question 2.1, one can write:

$$M = \rho V = \frac{4}{3}\rho\pi r^3.$$

By substituting this into the gravitational acceleration equation, it is clear that:

$$g_{Earth} = G\rho\frac{4}{3}\pi r.$$

Question 2.3: D

Repeating the same calculations done in the Question 2.2, but this time for the moon, we obtain:

$$g_{moon} = \frac{GM_{moon}}{R^2}.$$

Expressing the acceleration of the moon in terms of the density and radius of the earth (as in Question 2.1), one could write:

$$g_{moon} = \frac{G*\rho_{earth}*R_{earth}^3*\pi}{\frac{R_{earth}^2}{16}} = 16G\rho_{earth}R_{earth}\pi \implies g_{moon} = (3/16)g_{earth}$$

Question 2.4: A

For a satellite, we can equate the following:

$$F = mg = \frac{mv^2}{r}.$$

So, v is proportional to the square root of g – therefore, the speed will decrease with decreased g by a factor of $\sqrt{\left(\frac{3}{16}\right)}$.

Question 3.1: B

The resistance for resistors in parallel is the sum of the reciprocals:

$$\frac{1}{R_{total}} = \frac{1}{R} + \frac{1}{R} \implies R_{total} = \frac{R}{2}.$$

Question 3.2: E

Evaluate each group in parallel as one component and then add in series to first – remember to use 3.1 to make the calculation easier:

$$R_{total} = R + 1/\left(1/\left(R + 1/\left(\frac{1}{R} + \frac{1}{R}\right)\right) + 1/\left(R + 1/\left(\frac{1}{R} + \frac{1}{R}\right)\right)\right)$$

$$\Rightarrow R_{total} = R + 1/\left(2/\left(R + \frac{R}{2}\right)\right) = R + 1/\left(2/\left(\frac{3R}{2}\right)\right) = R + \frac{3R}{4} = \frac{7R}{4}.$$

Question 3.3: B

Using the equation for the voltage in the circuit, $V = IR_{total}$, one can use total resistance from previous question.

$$V = IR_{total} \Rightarrow R_{total} = \frac{V}{I} = \frac{1.4}{2} = 0.7 \ \Omega.$$

$$\Rightarrow R_{total} = \frac{7}{4}R \Rightarrow R = 0.4 \ \Omega.$$

Question 3.4: D

Using the equation for the power in terms of voltage and current $P = VI$, one can use the total current and voltage from question 3.1.

$$\Rightarrow P = 1.4 \times 2 = 2.8 \ W.$$

Question 4.1: B

Since the position is given by the equation $x = 10 + 1.5t^3$, and knowing that the velocity is a rate of change of the position, that is $v = \frac{dx}{dt}$, we have:

$$\Rightarrow v = \frac{dx}{dt} = \frac{d(10 + 1.5t^3)}{dt} = 4.5t^2$$

Question 4.2 D

The acceleration represents the rate of change of the velocity, so if one takes the derivative of the velocity in relation to time, one will have expression for acceleration.

$$a = \frac{dv}{dt} = \frac{d(4.5t^2)}{dt} = 9.0t.$$

Equating the acceleration to zero gives:

$$a = 9.0t^2 = 0 \Rightarrow t = 0.$$

Question 4.3 C

One can apply the equation for the position of the particle to figure out its total displacement for the given interval.

$$t = 2s \Rightarrow x_1 = 22 \ m, \ \ t = 10s \Rightarrow x_2 = 1510 \ m$$

$$\Rightarrow \text{Total displacement: } \Delta x = x_2 - x_1 = 1488 \ m$$

$$\therefore \text{Average velocity: } \Delta v = \frac{\Delta x}{\Delta t} = \frac{1488}{8} = 186 \ m/s.$$

Question 4.4 A

First step is to calculate the average velocity of the ball for the first 5 seconds.

$t = 5s \Rightarrow x = 197.5 \text{ m} \Rightarrow v = \frac{\Delta x}{\Delta t} = \frac{197.5}{5} = 39.5$ m/s.

Considering that an elastic collision is in place, the total energy transferred to the wall is equal to the kinetic of the ball in the moment of collision. Thus:

$E = \frac{mv^2}{2} = \frac{(m*39.5^2)}{2} \approx 780m$ J.

Question 5.1: C

The gravitational potential energy is given by the following:

$\Rightarrow E = mgh = (700 + 800) \times 10 \times 7 = 105,000$ J.

Question 5.2: B

This is a straightforward calculation:

Power = Energy / Time = 3500 W.

Question 5.3: A

The average velocity needs to be determined to find the kinetic energy:

Average velocity = Distance / Time $= 0.23$ ms^{-1}

$\Rightarrow$ Kinetic energy $= \frac{mv^2}{2} = 40.8$ J.

Question 5.4: C

The lift accelerates with the acceleration a for $t = 10$ s, then it uniformly moves for $t = 10$ s with the speed $v = at$ and, finally, it decelerates with the same acceleration for $t = 10$ s. Therefore:

$h = \frac{at^2}{2} + vt + \frac{at^2}{2} = 2at^2 = 2vt$

$\Rightarrow v = \frac{h}{2t} = 0.35$ ms^{-1}.

Question 6.1: E

One can apply Newton's second law to block 2 and solve the equation: $F_2 = m_2 a_2$. The component of the weight pulling the block down can be expressed as: $F_{x2} = m_2 g \sin\alpha$.

The component of the weight in the y direct (equals to the normal reaction against the ramp) can be expressed as: $F_{y2} = m_2 g \cos\alpha$. Therefore, the friction force can be expressed as: $F_{f1-2} = F_{y2}\mu_2 = m_2 g \cos\alpha \mu_2$.

Finally, applying Newton's second law, the acceleration of block 2 can be expressed as:

$m_2 a_2 = m_2 g \sin\alpha - \mu_2 m_2 g \cos\alpha \Rightarrow a_2 = g(\sin\alpha - \mu_2 \cos\alpha)$.

Question 6.2: A

Same reasoning applied in question 6.1 can be applied to block 1 to express the acceleration of block. The component of the weight pulling the block down can be expressed as: $F_{x1} = m_1 g \sin\alpha$.

The component of the weight in the y direct (equals to the normal reaction against the ramp) can be expressed as: $F_{y1} = m_1 g \cos\alpha$. Therefore, the friction force can be expressed as: $F_{f1-2} = F_{y1}\mu_1 = m_1 g \cos\alpha \mu_1$

Applying Newton's second law:

$$m_1 a_1 = m_1 g \sin\alpha - \mu_1 m_1 g \cos\alpha \Rightarrow a_1 = g(\sin\alpha - \mu_1 \cos\alpha).$$

Question 6.3: B

To determine the acceleration of m_1 in respect to m_2, one can simply subtract the acceleration of block 1 with respect to the plane from the acceleration of block 2 with respect to the plane:

$$a_1' = a_1 - a_2 = g(\sin\alpha - \mu_1\cos\alpha) - g(\sin\alpha - \mu_2\cos\alpha) = g\cos\alpha(\mu_2 - \mu_1).$$

Question 6.4: A

When the forces on block 2 are balanced, we can equate the following:

$$m_2 g \sin\alpha = \mu_2 m_2 g \cos\alpha$$

$$\Rightarrow \mu_2 = \sin\alpha/\cos\alpha = \tan\alpha.$$

Question 7.1 D

This question can be easily solved if the total time is divided in small intervals of time and analyzed individually. 30 min corresponds to 1800s. According to the graph:

Interval: 0 – 100 s: Accelerated movement: $a = \frac{\Delta v}{\Delta t} = \frac{20}{100} = 0.2$ m/s^2. Distance can be then calculated:

$$v^2 = v_o^2 + 2ad \Rightarrow d_{0-100} = \frac{v^2 - v_o^2}{2a} = \frac{20^2 - 0}{2*0.2} = 1000 \text{ m}.$$

Interval: 100 – 800 s: Constant velocity: $v = 20\frac{m}{s} \Rightarrow d_{100-800} = v * \Delta t = 20 * 700 = 14000$ m.

Interval 800 – 1000 s: Decelerated movement: $a = \frac{\Delta v}{\Delta t} = -\frac{20}{200} = -0.1\frac{m}{s^2}$. Distance can be

calculated: $v^2 = v_o^2 + 2ad \Rightarrow d_{800-1000} = \frac{v^2 - v_o^2}{2a} = \frac{0 - 20^2}{2*-0.1} = 2000$ m.

Interval 1000 – 1600 s: The velocity is zero which means that the car is not moving.

Interval 1600 – 1800 s: Accelerated movement: $a = \frac{\Delta v}{at} = \frac{15}{200} = 0.075$ m/s^2. Distance can be

calculated: $v^2 = v_o^2 + 2ad \Rightarrow d_{1600-1800} = \frac{15^2 - 0}{2*0.075} = 1500$ m.

Total Distance: $D = 1000 + 14,000 + 2000 + 1500 = 18,500$ m.

Question 7.2: D

Using information from previous question one can write the expression for the velocity in the interval 1600 − 1800 s. $v = v_0 + at = 0 + 0.075t$.

No calculation is needed for this problem if the student realizes that the integration of the velocity for that in the interval will give you the distanced travelled by the car in the interval, which has already been computed in the previous question and it 1500 m.

The integration then will be:

$$\int_{1600}^{1800} v\, dt = \int_{0}^{200} 0.075\, t\, dt = 0.0375 t^2 \big|_{0}^{200} = 0.0375(200^2) = 1500 \text{ m.}$$

Question 7.3: B

At t = 900s, the car is subjected to an acceleration of $|a| = 0.1 \text{ m/s}^2$. This is easily determined by the gradient of the velocity-time graph.

According to Newton's second law:

$F = ma \Rightarrow 1000 * 0.1 = 100N$.

Question 7.4: A

At t = 50s, the car is subjected to an acceleration of $|a| = 0.2 \text{ m/s}^2$, leading to an applied force of:

$F = ma = 1000 * 0.2 = 200 \text{ N}$.

The force applied to the car by the engine is constant throughout the whole interval. You can calculate the average work done by the engine as follows:

$W = Fd = 200 * 1000 = 200000 \text{ J}$.

Finally, one can calculate the power:

$P = \frac{W}{\Delta t} = \frac{200000}{100} = 2000 \text{ W}$.

Question 8.1: C

One can observe that the force applied to the particle is positive up to x = 3 m, which means that the particle is being accelerated up to that point, and its velocity is increasing.

After x = 3 m, the force applied to the particle starts to be negative, which means that the particle is being decelerated and its velocity decreasing after x = 3m. Therefore, the velocity is maximum at x = 3 m.

Question 8.2: D

Following similar reasoning as in the previous question, and knowing the maximum velocity happens at x = 3m, consequently, one can affirm the kinetic energy is maximum at x = 3 m as it is equal to ½ mv².

Question 8.3: A

One can get this answer by analysing the graph. The area under the curve of a graph (force x distance) will give the work that has been done to the particle. Since the particle is at rest at x = 0, its kinetic energy and velocity are zero at x = 0.

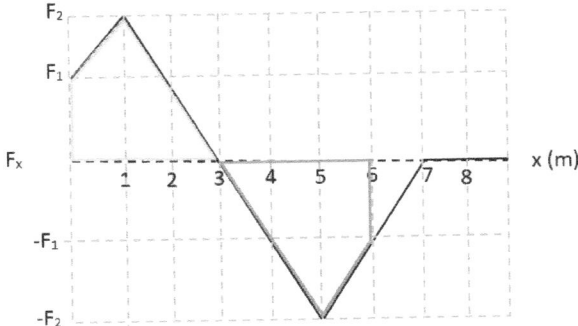

The area from x = 0 to 3 represents the work done to the particle by the positive force. The area from x = 3 to 6, represents the work done to the particle by the negative force; that is, the force in the opposite direction of the positive force.

Since the two areas cancel each other exactly at x = 6, all the work that has previous been done the particle by the positive force will have been cancelled by the work done by the negative force (acting in the opposite direction). Therefore, the kinetic energy of the particle at x = 0 is zero and, consequently, its velocity is also zero.

Question 8.4: D

After x = 6 metres, the force is acting in the opposite direction compared to at x = 2 m. Therefore, the acceleration vector is also pointing the opposite direction as it was in x = 2 m.

Question 9.1: C

This set of questions tests your knowledge of accelerated and decelerated motion and conservation of energy. When the ball is thrown up in the air, it is subjected to the gravitational acceleration, pointing downwards towards the centre of the Earth. Therefore, the acceleration is constant and negative, with the velocity reaching zero at the highest point and the same velocity, in modulus, when it returns to the starting point. Option C is the only one that represents such behaviour.

Question 9.2: B

To solve this question, one should consider the conservation of energy, since the resistance of air can be neglected.

Considering the position from which the ball leaves as h = 0, the potential energy of the ball will be zero. Since the ball leaves the girl's hand at v = 15m/s, its initial kinetic energy can be calculated:

$$E_k = \frac{1}{2}m.v^2 = \frac{1}{2} * 0.5 * 15^2 = 56.25 \text{ J}.$$

Considering conservation of energy and knowing that, at the highest position, the kinetic energy of the ball is zero, one can assume that the kinetic energy the ball had at the beginning of its trajectory will be entirely converted in potential energy. From this assumption, one can calculate the high reached by the ball:

$$E_k = E_p \Rightarrow 56.25 = mgh \Rightarrow h \approx 12 \text{ m}.$$

Question 9.3: A

One can assume this represents a case of decelerated movement with $a = -9.8 \text{ m/s}^2$, and easily calculate the time that takes for the ball's velocity be reduced from 15 m/s to 0:

$$v = v_0 + a.t \Rightarrow t = \frac{(v-v_0)}{a} = \frac{(0-15)}{-9.8} = 1.53 \text{ s}.$$

Question 9.4: C

First, on can calculate the potential energy of the ball 5.0m from the ground:

$$E_{P(5.0m)} = mgh = 0.5 * 9.8 * 5 = 24.5 \, J.$$

Since, there is conservation of energy, one can subtract the potential energy at 5.0 m from the kinetic energy of the ball at the starting point (calculated in question 9.1):

$$E_{k(5.0m)} = 56.25 - 24.5 = 31.75 \, J.$$

Question 10.1: C

Option B displays the correct representation of the forces with the weight of Object 2 pointing downwards towards the centre of Earth, the normal reaction against the surface of the ramp and friction force pointing downwards along the surface of the ramp as a reaction against the eminent movement of Object 2 upwards along surface of the ramp.

Question 10.2: E

Considering that Object 2 stays at rest in relation to the ramp, one can equate the vertical components of the forces acting on the object:

$$F_f \sin\theta + mg = N\cos\theta$$

$$mg = N\cos\theta - F_f\sin\theta = N\cos\theta - N\mu\sin\theta = N(\cos\theta - \mu\sin\theta).$$

Question 10.3: D

Following similar reasoning as in the previous question, one can equate the horizontal components of the forces acting on the object:

$$N\sin\theta + F_f\cos\theta = m_2 a$$

$$\Rightarrow m_2 a = N\sin\theta + N\mu\cos\theta = N(\sin\theta + \mu\cos\theta)$$

$$\Rightarrow a = \frac{N}{m_2}(\sin\theta + \mu\cos\theta).$$

Question 10.4: A

According to Newton's second law:

$$F = ma = (m_1 + m_2)a \Rightarrow F = 20a \Rightarrow a = \frac{F}{20}.$$

Dividing the answers for question 10.3 and 10.2 leads to:

$$\frac{a}{g} = \frac{(\sin\theta + \mu\cos\theta)}{\cos\theta - \mu\sin\theta}$$

$$\frac{F}{20g} = \frac{(\sin\theta + \mu\cos\theta)}{\cos\theta - \mu\sin\theta} \Rightarrow F = 20g * \frac{(\sin\theta + \mu\cos\theta)}{\cos\theta - \mu\sin\theta} = 219N.$$

FINAL ADVICE

Arrive well-rested, well-fed and well-hydrated

The ENGAA is an intensive test, so make sure you're ready for it. Ensure you get a good night's sleep before the exam (there is little point cramming) and don't miss breakfast. If you are taking water into the exam, make sure you've been to the toilet before so you don't have to leave during the exam. Make sure you're well rested and fed in order to be at your best!

Move on

If you find yourself struggling on a particular question, move on. Every question has equal weighting and there is no negative marking. In the time it takes to answer one hard question, you could gain three times the marks by answering the easier ones. Be smart to score points - especially in Section 2 where some questions are far easier than others.

Afterword

Remember that the route to a high score is a methodical approach and consistent practice. Do not fall into the trap that *"you can't prepare for the ENGAA"*– this could not be further from the truth. With knowledge of the test, some useful time-saving techniques, and plenty of practice, you can dramatically boost your score.

Work hard, never give up and do yourself justice.

Good luck!

Acknowledgements

I would like to express my sincerest thanks to the many people who helped make this book possible, especially the 15 Oxbridge Tutors who shared their expertise in compiling the huge number of questions and answers.

Rohan

SO YOU'VE MADE IT THIS FAR....

At the beginning, we told you to read on, enjoy the book, and master the exam – it's our hope that you've been able to do just that.

You've worked through practice questions, you've honed your techniques, and you feel more prepared than ever to face the exam. What's next?

The work has only just begun. You've prepared for the test, but that is only one part of your admissions journey. For the very brightest students, like yourself – competition becomes the most rife during the interview stage, and every year so many students ace the test only to fall at the last hurdle.

At UniAdmissions we care about guaranteeing that everyone we work with has as much support as they need *every step of the way*, and we have your back.

The Ultimate Oxbridge Interview Guide and the Ultimate Oxbridge Interview Guide: Physical Science are exclusive titles designed to help you master the interview like you've mastered the ENGAA.

Chloe Bowman, Vendy Fialkova
Matthew Fox & Dr Rohan Agarwal

THE ULTIMATE OXBRIDGE INTERVIEW GUIDE – PHYSICAL SCIENCE

Find them online through Amazon, WHSmiths, and Waterstones.

The Ultimate Oxbridge Interview Guide: Phys. Science - https://amzn.to/3DDmYp8

- Learn from HUNDREDS of real interview questions with FULLY WORKED SOLUTIONS
- Comprehensive overview of all related subjects for maximum coverage
- You can't get this prep anywhere else

The Ultimate Oxbridge Interview Guide - https://amzn.to/3J53QkP

- Prepare for a wide range of interview techniques
- Work through hundreds of practice questions
- Learn tips and tricks from real interviewers and examiners for top UK medical schools

About Us

We currently publish over 85 titles across a range of subject areas – covering specialised admissions tests, examination techniques, personal statement guides, plus everything else you need to improve your chances of getting onto competitive courses such as medicine and law, as well as into universities such as Oxford and Cambridge.

This company was founded in 2013 by Dr Rohan Agarwal and Dr David Salt, both Cambridge Medical graduates with several years of tutoring experience. Since then, every year, hundreds of applicants and schools work with us on our programmes. Through the programmes we offer, we deliver expert tuition, exclusive course places, online courses, best-selling textbooks and much more.

With a team of over 1,000 Oxbridge tutors and a proven track record, UniAdmissions have quickly become the UK's number one admissions company.

Visit and engage with us at:

Website (UniAdmissions): www.uniadmissions.co.uk

Facebook: www.facebook.com/uniadmissionsuk

YOUR FREE BOOK

Thanks for purchasing this Ultimate Guide Book. Readers like you have the power to make or break a book —hopefully you found this one useful and informative. *UniAdmissions* would love to hear about your experiences with this book. As thanks for your time, we'll send you another ebook from our Ultimate Guide series absolutely <u>FREE</u>!

How to Redeem Your Free Ebook:

1) Either scan the QR code or find the book you have on your Amazon purchase history, or your email receipt, to help find the book on Amazon.

2) On the product page at the Customer Reviews area, click 'Write a customer review'. Write your review and post it! Copy the review page or take a screen shot of the review you have left.

3) Head over to www.uniadmissions.co.uk/free-book and select your chosen free ebook! Your ebook will then be emailed to you – it's as simple as that!

Alternatively, you can buy all the titles at:

www.uniadmissions.co.uk

Printed and bound by CPI Group (UK) Ltd, Croydon, CR0 4YY

22/04/2026

02094796-0001